T0261116

STEPPING IN THE SAME RIVER TWICE

EDITED BY AYELET SHAVIT AND

AARON M. ELLISON

WITH A FOREWORD BY W. JOHN KRESS

Stepping in the Same River Twice

REPLICATION IN BIOLOGICAL RESEARCH

Yale UNIVERSITY PRESS   NEW HAVEN AND LONDON

This volume is a contribution of the Long Term Ecological Research (LTER) program funded by the U.S. National Science Foundation. The twenty-five LTER sites, which encompass diverse ecosystems from Alaska and Antarctica and include islands in the Caribbean and the South Pacific, make up the largest and longest-running ecological research network in the world. The LTER network of sites serves as a global model for transparent and reproducible ecological and environmental research.

Yale University Press books may be purchased in quantity for educational, business, or promotional use. For information, please e-mail sales.press@yale.edu (U.S. office) or sales@yaleup.co.uk (U.K. office).

Set in Scala and Scala Sans type by Westchester Publishing Services. Printed in the United States of America.

Library of Congress Control Number: 2016951790
ISBN 978-0-300-20954-9 (hardcover : alk. paper)

A catalogue record for this book is available from the British Library.

This paper meets the requirements of ANSI/NISO Z39.48-1992 (Permanence of Paper).

10 9 8 7 6 5 4 3 2 1

To Ran, Ofri, Tal, Rotem, and Frodo,
who remain unique and unreproducible,
and
To the past, present, and future practitioners of Open Science

History does not always *repeat* itself. Sometimes it just yells, "Can't you remember anything I told you?" and lets fly with a club.

—*John W. Campbell, from* Analog Science Fiction and Science Fact *(1965)*

CONTENTS

W. John Kress

Here is the idealized scientific method at work: (1) a testable hypothesis is constructed to determine a general principle that explains some initial observations; (2) an experiment is devised that will generate data to test the hypothesis; (3) the experiment is conducted and data are collected; (4) the data are analyzed; (5) the experiment is repeated until sufficient evidence allows one to fully accept or reject the hypothesis that explains the original observations; and (6) the accepted hypothesis may result in new insights that require additional testing. This basic method enables scientific progress to be made over time. In some cases the initial hypotheses can be narrow; in other cases they can be broad and substantial. In all cases the experiments, the data, and the analyses need to be replicable, repeatable, and reproducible. Good science that follows this basic method enables us to understand the world around us.

In today's society, in which science is responsible for all aspects of our lives, from our understanding and utilization of nature to our health, to our economies and livelihoods, and to our national security, scrupulously adhering to this scientific method is essential for civilization to progress and for human populations to prosper. Yet in the first decades of the twenty-first century, the integrity of this scientific process, and of the scientists who depend on it, has in some cases and by some individuals been called into question. Are the scientific method and the reproducibility of scientific results being compromised in the highly competitive nature of today's world?

The question of reproducibility is the topic of *Stepping in the Same River Twice: Replication in Biological Research*, superbly edited by Ayelet Shavit and Aaron M. Ellison, a philosopher of science and a field ecologist, respectively. The editors and other contributors to this volume provide a broad spectrum of thoughts and opinions on the topic. It is a subject highly relevant to the role of science in society.

First of all, as clearly presented in the book, replication, replicability, repeatability, and reproducibility must be defined. Replication is a general term, denoting different concepts and practices. A replicate results from the duplication of the methods, the results, and the analyses of an experiment at the same timeframe as the original procedures are performed. There is no variance, no error, and no secondary interpretation. Everything is the same. Repeatability is the repetition of the original experimental process under similar, but not exact, laboratory, field, and environmental conditions, which in essence tests the original conditions of the experiment under new and controlled settings. Everything is similar but not quite the same. Reproducibility is the attempt to obtain the same results of an experiment at a different time and in a different space. Exact reproducibility is difficult to achieve and for the most part is infrequently attempted from one lab to the next.

The notions of replication, repeatability, and reproducibility are necessary to successfully engage in the scientific method. Replicates are vital to reducing experimental error. Repeatability is core to understanding the laws of nature, as they allow us to quantify how processes and features, such as evolution, ecology, physiology, and morphology, vary from place to place, from one point in time to the next, and from one individual to a population of individuals, within and between species. Reproducibility allows scientists to question why experiments differ from lab to lab and between environments and therefore serves as a test of the credibility of our results.

However, it would be a big mistake to equate the lack of reproducibility with poorly conducted science or even fraud. Repeatability is simply the basic nature of the business of science. Exact reproducibility is in some ways the pipe dream of a scientifically illiterate society looking for absolute truth in what it does not understand. For scientists, and for those in society who understand the scientific process, science is working well when irreproducible results and conclusions are revealed, when they are not repeatable, and when new insights about nature are the result. It is critical that we distinguish between experimental and observational error, natural variability in environmental conditions, and fraudulent misconduct in the scientific method.

The execution of the scientific method requires the creditable testing of initial hypotheses, it requires the necessary replication of earlier experiments, and it requires obligatory trials to exactly reproduce original results. The scientific method is the opposite of fraud and misconduct. In the popular and the scientific media, terms such as reproducibility, transparency, confidence, rigor, independent verification, blind analyses, openness, open-source, strict guidelines, preclinical studies, incentives, fraud, misconduct, and retraction are now common (see McNutt 2014a, 2014b; Buck 2015; MacCoun and Perlmutter 2015). Unfortunately, often what is implicit in this media coverage is that the latter actions, namely "fraud, misconduct, and retraction" are becoming the norm in our scientific endeavors. This conjecture could not be further from the truth.

For example, a recent article about scientific integrity in *The Economist*, entitled "Unreliable Research: Trouble at the Lab" (19 October 2013), lacks any explanation of the scientific process, equates the lack of reproducibility with misconduct, and for all practical purposes addresses only the biomedical sciences, completely ignoring the fields of basic biology, ecology, physics, chemistry, mathematics, and so on. The field of biomedicine is indeed critical to the advancement of a prosperous society but in some ways is more a combination of science, industry, and investment than a true science itself.

The scientific community, including authors in the present volume, has mounted a serious response to the issue of reproducibility in science. Within the major scientific journals and professional societies discussion and debate have centered on the extent of irreproducibility in scientific research and what should be done about it. Significant and far-reaching changes have been devised and introduced in the areas of scientific funding, publication, and research institutions. Some of the proposed ideas are hot off the presses and already have been given acronyms, such as conflict of interest (COI), good institutions of practice (GIP), and transparency and openness promotion (TOP).

COI has become a critical issue for funding agencies around the world, including the National Institutes of Health and the National Science Foundation in the United States. The assurance that scientists are not unduly influenced by their funders, including corporate, foundation, and philanthropic supporters, is a major concern on all fronts. Conflict-of-interest statements are now required by many journals and funding bodies to make sure that science is conducted in an objective fashion unbiased by sources of financial support, yet little consensus exists on the exact nature of what scientists should disclose (see Reardon 2015).

GIP is an idea that has been proposed as a set of guidelines for research institutes to make sure that basic laboratory environments for conducting science are conducive to the scientific process and encourage scientists to focus on robust results rather than rapidly generated, high-profile outcomes. These institutional practices should include routine discussion of research methods among scientists, should establish open reporting systems, should emphasize training of students in transparent scientific practices, should make available notebooks and laboratory records for independent review, and should devise appropriate incentives and evaluation systems for a responsible scientific community (see Begley, Buchan, and Dirnagl 2015). Although none of these guidelines are mandatory, they define ideals for increasing the proper conduct of science at major research institutions.

Finally, TOP is becoming a standard for scientific journals, especially in the field of biomedicine, to encourage transparency, openness, and reproducibility in research and publication (see Nosek et al. 2015). These recently adopted guidelines have set-up standards and levels of compliance as incentives for the publication of sound and reproducible science in high-profile journals. These standards include (1) open practices in citations and replication protocols; (2) definitions for openness in design, research materials, data, and analytic methods; and (3) pre-registration procedures for establishing beforehand the purpose and scope of the publications. TOP is meant to insure reproducibility and honesty not only in scientific publications but in the conduct of science itself.

However, are these new guidelines and requirements being insti-
tuted by funding agencies, research institutions, and scientific journals
fundamental and essential to a new era of better and reproducible science? Or are they simply over-reactions by our community to address and
alleviate the concerns of the public about the legitimacy of science itself?
Reproducibility has always been fundamental to good science. But the
recent preoccupation with irreproducibility may be a reflection of the
broader societal issue of overall credibility and integrity and not simply
a problem restricted to the scientific method.

No matter how hard some journalists and editors in their commen-
taries attempt to address the lack of transparency in science, much of
the current news on irreproducibility in science equates irreproducible
results with fraudulent behavior, especially in the high-profile, for-profit
biomedical industries (see Alberts et al. 2015). This situation is simply
not the case for the larger discipline of science as a whole. Misconduct
and fraud are not overwhelming problems for science. What is overwhelming is the prevalence of misconduct and fraud in all sectors
of today's society, including industry, commerce, banking, the market-
place, the military-industrial complex, politics, and governments, not
just in science. If we are to truly understand the importance of reproducibility in the sciences, we will need to first address the general prevalence of misconduct in our society and the causes behind it. Simply
attacking and trying to correct the field of scientific inquiry with new
rules and standards may be a misguided solution that misses a more
deeply rooted problem embedded in today's world as a whole. Understanding the fundamental role and necessity of reproducibility and
repeatability in the scientific method itself is a necessary first step in the
process. This book provides an informative roadmap to that end.

Doing science is simultaneously very easy and unfathomably difficult. Our senses are bombarded constantly with information—sights, sounds, smells, tastes, and tactile sensations—about the world around us that we rapidly and often unconsciously turn into ideas about how the world works. Scientists simply take these ideas and test them using a combination of observations, experiments, statistics, and models. Anyone can make observations, run experiments, and analyze the data, but translating the results into successful scientific theories requires at least one additional step: confirmation that the results can be repeated in the same or different laboratories or field sites and that the analyses can be reproduced independently.

Successful replication of scientific findings turns out to be surprisingly hard. In the past few years, scientific journals and the popular media have been replete with papers and stories claiming that fewer than 10 percent of scientific studies in fields ranging from cancer treatment and neurobiology to ecology and evolutionary biology are reproducible. These same papers and stories each focus on particular reasons for the lack of replicability, such as small sample sizes, insufficient methodological details, unavailability of data or computer code, or outright fraud. They each then propose unique solutions: larger sample sizes, extensive (and expensive-to-publish) methods sections for every paper, massive data archives that adhere to universal metadata standards, statistical editors, and offices of scientific integrity. Individually, each of these solutions has appeal, but in aggregate they still fail to grapple with the underlying issue of why the replication of scientific findings should be so hard. This book is our attempt to address this deeper issue.

We (the co-editors) first met in 2007 at the University of California, Davis, where we discovered that a philosopher of science and a field ecologist-cum-statistician shared a common interest in the problem of successfully revisiting historic research questions, repeating landmark studies, and monitoring ongoing environmental change. But we also found that we lacked a common language. What seemed like an easy question to an ecologist—when and where a study was done—could vex a philosopher wrestling with the concept of time and location. As we continued for a few more years to discuss the challenge of repeating a study in time and space—Chapter 1 summarizes our years of work toward a common language to use in describing replication—the question of reproducibility of scientific findings started making headlines. We rapidly realized that our philosophical questions about replication and reproducibility far transcended ecology and that addressing them might help bring clarity to ongoing debates across the sciences.

Scientists have not been alone in their quest to understand reproducibility. In Chapters 2 and 3, Yemima Ben-Menahem and Haim Goren, respectively, illustrate how writers, philosophers, and geographers have conceptualized replication in the service of memory, cartography, and religion. These philosophical and historical lessons remind us that replication, repetition, and reproducibility concern everyone—not only scientists—because these processes define our interactions with the world around us.

The broad, conceptual ideas discussed in the first three chapters are brought back down to earth through a series of discipline-specific overviews and case studies. Each chapter is presented as a dialogue in two parts: first, theory and background; second, a detailed case study. These include discussion of the importance of museum specimens and samples (Chapter 4 by Tamar Dayan and Bella Galil) paired with a case study of using museum specimens to guide modern resurveys to look for signals of ongoing environmental change (Chapter 5 by Rebecca J. Rowe); the structure and implementation of a national program to monitor environmental change (Chapter 6 by Avi Perevolotsky and his colleagues) on land (Chapter 7 by Ron Drori and his colleagues) and at sea (Chapter 8 by Jonathan Shaked and Amatzia Genin); experimental design and statistical analysis (Chapter 9 by Aaron M. Ellison) and their importance in uncovering and accounting for variability among individuals in immunological research (Chapter 10 by Jacob Pitcovski and his colleagues); how to combine results of studies using systematic reviews and meta-analysis (Chapter 11 by Leonard Leibovici and Mical Paul) and their application to use of antibiotics to treat cholera (Chapter 12 by Mical Paul and her colleagues); and the value and importance of metadata in creating trustworthy science (Chapter 13 by Kristin Vanderbilt and David Blankman), together with the challenges in producing the metadata scientists and decision makers actually need (Chapter 14 by Emery R. Boose and Barbara S. Lerner).

We conclude with three chapters and an epilogue synthesizing the ideas and key points raised by the overviews and case studies. Morgan W. Tingley (Chapter 15) focuses on ensuring that comparisons between studies that are meant to replicate one another really do and discusses how hierarchical models can be used to combine data from disparate yet comparable sources into broader syntheses. Barbara Helm and Ayelet Shavit (Chapter 16) return to the challenge of identifying time and location and remind us that human time and space are not the same as time and space for many, if not all, of the organisms that biologists study. Replication depends critically on maintaining an accurate frame of reference for time and location; doing so takes us far away from hours, minutes, or seconds and degrees of latitude and longitude. Aaron M. Ellison (Chapter 17) and Ayelet Shavit (Epilogue) suggest a series of best practices for creating replicable research that cross disciplinary boundaries. Many of these best practices individually recall solutions proposed in the hundreds of articles published on this reproducibility in the past several years, but having them all in one place, grounded by the reality of the case studies, should further facilitate the production of reliable and trustworthy open science.

ACKNOWLEDGMENTS

This book is like a chorus, comprising many different voices working in a mutual and coordinated effort to bring about harmony in thought and practice. Three years of repeated encounters and discussions across three continents took place between the participating biologists, philosophers, and historians. Their effort would not have materialized into this more-than-the-sum-of-its-parts text without the effort of many others, who we deeply appreciate and thankfully acknowledge. We especially thank Jean Thomson Black, senior executive editor at Yale University Press, who from the start believed in our project and supported the publication of this book from its very first step. Samantha Ostrowski at Yale University Press kept us on track and obtained four very useful anonymous reviews; these, along with readings of the entire manuscript by David Foster (Harvard Forest), Julia Jones (Oregon State University), and Eva Jablonka (Tel Aviv University), greatly helped us improve the book you are now reading.

This book was launched at the Ecological Society of America's 2013 Annual Meeting in Minneapolis, Minnesota, and we were delighted that the Program Committee accepted the symposium proposal that brought many of the authors together for the first time. We are grateful to all the smart and dedicated people who helped organize the follow-up workshop in 2014 at Tel Hai College: the students Bar Katin, Tomer Rotestein, Iris

Eshed, Yael Silver, Sivan Margalit, and Rieke Kedoshim, and administrator Dianna Rachamim. We particularly thank the President of Tel Hai College, Yona Hen, for gambling on the workshop and advancing the funds to support it. Once the book began to take form, we were fortunate to have the assistance of very talented artists: Tamar Ben Baruch, who prepared the flowchart in the Epilogue, Dror Her Miller, who granted permission to use his painting, Padawan Anat Kolombus for careful technical editing and indexing of the manuscript, and the production team at Yale University Press for outstanding work. Finally, we thank the Israel Science Foundation (grant numbers 960/12 and 960/14), the US National Science Foundation (grant numbers 0620443 and 1237491 for the Harvard Forest Long Term Ecological Research program), Tel Hai College (grant number 6/15), and the Harvard Forest for financial and logistical support throughout the project. To all these organizations and people we send our deep appreciation, and we apologize to anyone we have inadvertently overlooked. To our friends and families, who may never read this book but who allowed us the time and pleasure to contemplate, with neither guilt nor regret, the ethereal scientific and philosophical concepts discussed herein, we don't really know what to say. Words end and wonder enters. Thank you.

PART ONE INTRODUCTION
REPLICATION ACROSS DISCIPLINES

## CHAPTER ONE

Toward a Taxonomy of Scientific Replication

Ayelet Shavit and Aaron M. Ellison

*"Replication" encompasses many different meanings and denotes
a range of goals and motivations in diverse disciplines. Explicating
and clarifying these meanings will help to bridge gaps between
disciplines and strengthen interdisciplinary understanding.*

Society relies on scientific theories and information to manage virtually
all aspects of civilization, and the technologies we use every day are the
visible manifestations of science. Replication is fundamental to the development and application of science. Without replication, the value and
trustworthiness of scientific results and theories remain in doubt. But
replication is also like a pair of eyeglasses: it is typically ignored but is
nonetheless indispensable. One major motivation for this book is to increase the trustworthiness of science by critically examining how scientists, especially biologists, replicate their results. A mixture of theoretical
development, synthesis, and case studies also reveals ways to improve
the processes of replication.

Replication is "the set of technologies which transforms what counts
as belief into what counts as knowledge."[1] Replication is especially important because it allows scientists to build general patterns and theories
from specific case studies.[2] The derivation and explication of general

theories are crucial for science. Without general theories, we could not make sense of natural phenomena or trust predictions of natural laws. In this first chapter, we suggest a taxonomy of the concept of replication that can be used to build bridges among different research cultures in the biological sciences—lab or field-based; observational or experimental; ecological, agricultural, medical, or physiological—while minding the gaps between these cultures.

In common usage, "replication" seems to be a straightforward term that simply means "doing the same thing again." As a reader or researcher, you might well ask, What additional conceptual clarification could possibly be needed for understanding replication? In response, we invite you to imagine this situation: a group of sailors sets sail from the Port of Aden. Before leaving, the captain orders: "From now on, and throughout the rest of our journey, you need to constantly circle the main mast in the same direction." "In what direction?" they ask. "In the direction that the water flows," comes the answer. Day and night the ship sails south, with the devoted sailors circling the mast with the water flowing around the ship from starboard to port. After a few weeks, however, after sailing south of the Equator and nearing the island of Zanzibar, the sailors observe that the sea now flows around the ship from port to starboard. The confused sailors ask the captain, "In what direction do we circle the mast now?" Their captain's response is swift: "What kind of a question is that? I already told you to do the same thing again!"

Replication of the captain's order is not clear in this simple imaginary example because the concept of replication implies radically different uses and goals simultaneously. If the sailors find "replication" ambiguous in this case, is it likely to be any easier for biologists or philosophers of science to use? Because it is not, we urge you to read this chapter carefully. We use it to sketch a map to be used while navigating the rest of the book. We delineate different ways that people use "replication"; point to specific chapters that explicate problems for replication encountered in different disciplines; and identify different models, practices, and heuristics to resolve them.

Heraclitus was perhaps the first person to identify the challenge of replication when he wrote, more than 2,000 years ago: "On those entering the same river twice other and other waters flow."[3] A millennium later, at the birth of modern empirical science during the seventeenth century, Robert Boyle was the first to write detailed guidelines to ensure that replicates of his vacuum chamber could be built so that his experimental results could be reproduced.[4] Concurrently, Francesco Redi extensively used replicates within each of his own experiments to, among other things, refute the theory of spontaneous generation.[5] Like their Medieval and early Modern science predecessors, seventeenth-century scientists aimed for exact and absolute replication—a perfect return to the original and "pure" source.

The demand for absolute replication diminished during the mid-nineteenth century, when science began to emphasize progress, new data, and innovative technology. If moving forward is one's scientific ideal, it is difficult to regard an exact return to one's point of origin as fulfilling that ideal. Tracking the turning point from such "absolute" replication to "progressive" scientific measurements is captured in two historical case studies of cartography concerning Jerusalem and the Dead Sea (see Chapter 3). In fact, all of the case studies in this book accept that repetition need be only "similar enough" rather than identical to previous results.

Given the shift in emphasis from science as exact copy to science as progress, it may not be surprising that replication is rarely, if ever, attempted for most published research in the life sciences and social sciences, and that some of that work may not even be repeatable or generalizable.[6] However, surprise aside, the lack of replication is clearly a current problem for science and of grave concern to scientists. The "problem of replication" is not only one thing; we first need to distinguish between the problem of detecting fraud or bad science and what this volume is really about, namely, how to identify, understand, and minimize the inherent difficulties in *any* study of a biological system.[7] High-quality biological science attempts replication with three different goals: (a) casually tracking a biological process;[8] (b) statistically validating the generality of a given study; and (c) empirically corroborating its accuracy. Each of these concepts, and tensions inherent in them, is dealt with extensively in this book.

Studies and results of research on biological systems almost always are contingent on the place and time at (or over) which observations are made or experiments are done.[9] The tremendous variability in scope, scale, data structure, and semantics involved in the study of any living system suggests that the problem of replication extends far beyond studies of any particular biological (sub)discipline.[10] These specific challenges are further examined and tackled in this book in a series of case studies on subjects ranging from cells (Chapter 10) to ecosystems (Chapters 5, 8); involving observations (Chapters 7, 8), experiments (Chapters 10, 16), and meta-analyses (Chapter 12). Appropriate analysis requires the careful crafting of metadata (Chapters 13, 14) and identification of relevant spatio-temporal scales by applying statistical tools such as occupancy modeling (Chapter 15) and autocorrelation functions (Chapter 16) to address the limits of spatial and temporal variability. Replication is a universal process that occurs in a range of settings. We explore a non-exclusive range of different contexts that set the stage where replication takes place: human clinics (Chapter 12); animal laboratories focused on agriculture (Chapters 10) or physiology (Chapter 16); natural history museums (Chapters 5, 15); and ecosystems ranging from the open ocean to temperate forests to extreme deserts (Chapters 4–8).

To be useful, the concept of replication in the biological sciences needs to be sufficiently clear and broadly applicable and also needs to provide specific-enough tools to allow scientists to answer three crucial questions for any scientific endeavor: To what degree are the changes in biological conditions tracked? To what degree are the scientific results accurate? And how generalizable are they? Answering the first question depends on monitoring and diagnosing one or more causal processes. Answering the second requires knowing whether observed results correspond to, and can be understood plausibly as, either a causal chain or a random process.[11] Answers to the third refer to the degree to which a datum collected from a given population—whether random or not—is representative of other populations.[12] In other words, the first question asks if change was consecutively tracked through time, the second question asks about the accuracy or realism of a scientific description and analysis (what actually happened in and at a given place and time), and the third question asks about its general representativeness and its range of possible predictions. These three questions are not the same—in some contexts they even conflict—yet all are important components of replication that weave through the ideas and case studies found throughout this book.[13] In the next section we more precisely define and characterize the concept of replication and discuss its links with generalizable theories. Through this presentation of a taxonomy of replication, we identify the range of possible practical meanings of the concept of replication that pertain to the goals, working stages, and common practices of scientists and scientific research cultures.

REPLICATION DEFINED

The terms "replicate" and "replication" are used not only in science but also in the arts, music, and law, and, we suspect, in everyday common usage, either as a verb representing the act of repetition or recapitulation or as a noun representing an instance of one of these.[14] The concept of replication in the English language is traceable first to Chaucer's discussion of language itself. Replication then emerges elsewhere: in architecture, a replica is a copy of a material object, typically on a smaller scale than the original object and built using different materials and processes, whereas in music a replica is a copy of a tone on a higher or lower scale of octaves than that of a given tone. There is a similar diversity of meanings of replication in the sciences. We identify the term "replication" as used in the life sciences to be a generic (higher-order) term that encompasses both more specific objects (nouns) and actions (verbs). In particular, we distinguish three more specific terms within the generic concept of replication: a "replicate" (or replica) within one's measurement process, "repeatability" (or repetition) between measurement processes, and "reproducibility" (or reproduction) between results of different measurements; we discuss each in more detail below.[15] All three reflect the goal of identifying the range of variation among individual measurements or individual pieces of information, which rarely are identical due to the complexity of living organisms or the underlying architecture of computers.

Wherever possible throughout this and subsequent chapters, we use one of these three more specific terms, but we use "replication" when more than one of these specific terms is relevant to a case study or discussion at hand or if it is not apparent which more specific term would best be applied in a particular case.

A *replicate* is a copy of a result—that is, of a piece of information or material object—that is produced during a single measurement process (using the same tools in the same manner and order) as the initial (replicated) result and on the same spatio-temporal scale. Scientific measurements are made either within or between observational and experimental sampling processes, but only a repetition within a measurement process is considered a replicate (or replica). For example, one may replicate a sampling event by collecting a second 100-ml bottle of seawater at the same longitude, latitude, and depth and within the same hour as the first (see also Chapters 6–8). One may also replicate a calibration event, for example, by taking a second temperature measurement of water in an individual water bottle, using the same thermometer that was used to make the first measurement.[16] Note that the actual temporal and spatial boundaries for any such replicates differ among organisms and research (sub)disciplines, and precision (or resolution) is context dependent. Thus, an oceanographer might insist that a replicate measurement of a water sample be taken within an hour of the first, whereas a marine geologist would accept a replicate sample of calcium concentration taken at an identical location regardless of elapsed time (Chapter 8).[17] Yet in both of these examples, as in many laboratory-based research fields, replicating the protocol is necessary, but not sufficient, to yield a replicate sample.

In contrast, in clinical trials (Chapters 11, 12), experiments in immunology (Chapter 10) or physiology (Chapter 16), and observational systematics or biodiversity surveys (Chapters 4, 5, 8, 15), the primary focus is on individual objects: a certain micro-array, blood sample, or specimen that was collected at one time and needs to be re-examined at another. True replicates do not, by definition, exist for individual organisms; yet in museum collections, for example, a species is often sorted and catalogued by its type specimen, a single or small group of individuals representing all other members of the species.[18]

In research collections held by museums, the "type specimen" is the object to which all other members of a species are compared, and the other members of the population normally are similar to (but not exactly the same as) the type specimen. Although variability among replicates within a species is expected, when an individual is determined to be substantially different (in physical form or genetic composition) from the type specimen, it may be considered to be a different species (e.g., in taxonomy; see Chapter 4), of different ancestry or race (e.g., in clinical trials; see Chapters 10, 12). At the same time, natural history collections and genome databases define the boundaries of variation for particular species and populations.[19]

Repeatability is the recurrence of a spatio-temporally distinct process of measurement—e.g., an experiment, survey, field-collection event, analytical model, or computer simulation "run"—done under similar conditions and usually with several replicates within each measurement process. Although most of the physical, environmental, laboratory, or field conditions under which repeated samples are taken will be similar (e.g., the equipment used, the type of organism observed), either the time(s), the location(s), or both will differ from one measurement process to another.

A repeated (second) measurement or result should be expected to be not the same as (or identical to) the first but only comparable to it.[20] Repeated measurements of additional samples can be used to quantify the range of variation among individual measurements.[21] Nonetheless, a study is considered to be repeatable if one is able to get comparable results—not necessarily identical ones but ones not differing beyond a reasonable doubt or outside of the bounds of expected variation—after repeating a new and relatively similar process of empirical measurement. For details on how to model such a comparison, see Chapter 15.

Finally, *reproducibility* refers exclusively to duplication of the results of a given study. A study is considered to be reproducible if one is able to get the exact same results after repeating the analysis using the same data at a different time but without making any additional empirical measurements.[22] In practice, exact reproducibility is difficult to attain, and one needs a statistical or quantitative method to model the observed range of variance and its departure from previous results. True reproducibility rarely is attempted in science, in part because scientific journals typically do not publish research papers focused on reproducibility and in part because the original data and metadata often are not transparent and accessible in ways that can enable the results to be reproduced even if one wanted to (see Chapters 13 and 14 for a short overview of the metadata problem and a detailed case study presenting a tool to use as a first step to resolving this problem).

PHASES OF RESEARCH AND TYPES OF REPLICATION

Replication is not one type of "thing" in either its definition or its application. The challenges that scientists face in doing replicable research vary with the different phases of their research project. We have identified five different phases of a project, and in each phase a different type of replication tends to overshadow the practice. These phases of a scientific project include the following:

1. Repeating an empirical measurement in the lab or the field (Chapters 4–10, 16, 17).
2. Creating comparable, analyzable, and generalizable data sets by examining and controlling the raw data from these repeated measurements (Chapter 15).
3. Filling in missing values and spatial or temporal gaps in a data set that combines data items from different data sources (e.g., field notebooks, spreadsheets, digital entries, or items streamed from remote sensors) and describing the combination process (Chapter 14).
4. Retrieving individual entities (e.g., data records, specimens, tissue samples, or other material objects) from an earlier stage of a study (Chapters 4, 5).
5. Repeating the analysis of the synthesized data or using meta-analysis to synthesize results from more than one comparable data set (Chapters 11, 12).

Each of these phases may not occur in every research project, nor is their order fixed. Keeping these phases in mind when planning a research project should help researchers to avoid many of the difficulties they encounter in ensuring that their results and analyses can be replicated; a flowchart for the benefits of such planning is presented in the epilogue (Chapter 18). Further, accomplishing any of these five different phases of research facilitates further generalization and aids in the development of scientific theories. We note that "generalization" also must be refined and clarified; we discuss generalization in this chapter in more detail after additional discussion of the different kinds of replication in the life cycle of a research project. Finally, identifying ties between replication and generalization sets the importance of replication into sharper relief.

Repeating Measurements

Under normal research conditions, a repetition of an observation or an experiment occurs at a time that is different from when the original observation was made or experiment run.[23] Re-surveys of historical studies, for which longitude, latitude, and elevation or depth are poorly documented present obvious challenges (see Chapters 5 and 15 for analytical ways to overcome and even benefit from this challenge). It may seem that the availability of high-precision global positioning systems (GPSs) should make repetition of contemporary studies at identical physical locations much more straightforward. But even when GPS locations are known precisely and the original researcher is still available to provide further guidance, it can be difficult to repeat an observation at the same location when there is more than one plausible way to link the research question and the procedure for identifying a location (see Chapters 5, 15, 16).

Monitoring programs with fixed locations can facilitate replication even when research questions have changed (Chapters 6–8). In some cases an expansion of the spatio-temporal scale of a replicate sample (i.e. coarse-scale re-sampling; Chapter 5) counter-intuitively may increase the accuracy of the overall results. There remain other cases, however, for which repeating an observation or an experiment at the same location turns out to be surprisingly difficult, even with cutting-edge technology (see Chapter 15 on how to use detectability and occupancy modeling to help tackle such dilemmas).[24] Finally, a more nuanced understanding of the variability in biological time (i.e., chronobiology) and space also is needed to improve experimental repeatability to better capture the casual dynamics of evolutionary and long-term environmental process (see also Chapter 16). To conclude, reliably repeating a measurement process is so basic for trusting scientific theories, so diverse—given the richness of the biological world and our ways of studying it—and so dynamic—given the constant flow of new technology—that any individual can only hope to do his or her best while minding and reporting any challenges or apparent errors (see Chapter 17 for a summary of best practices for repeating experiments).

Creating Reproducible Data Sets

The second and third kinds of replication involve tracking and marking not genes, birds, or water currents but data and data sets. Raw data records, in analogue or digital form, must be checked, entered into spreadsheets or databases, checked again, aggregated, and so on, to produce a data set: a synthetic product derived from numerous small decisions made by individuals and computers that (re)shape one or more sets or streams of raw data. Some of these decisions, such as removing individual entries that are unreadable in notebooks or out of range because of faulty sensors, derive from standard quality control and quality assurance norms and standards.[25] Others assess and analyze the data for their comparability with previous data collected under similar circumstances (see Chapters 13 and 14 for a discussion of how careful construction of metadata creates new possibilities for reproducibility).

Different researchers working at different times use different methods, tools, and techniques, and often different units of measurement as well. How can such data be reconciled and synthesized? One approach has been to develop a specific method or metric and accept only data that have been collected and reported as specified in formal protocols or codes. Many natural history and biodiversity collections, large location-based instruments such as telescopes and particle accelerators, repositories for DNA sequence data, and emerging national and international networks of environmental observatories all have such codes and standard methods for producing acceptable data.[26] This approach promotes the ideal of general representativeness over fine-tuned accuracy, which seems reasonable given the global scale of many of these organizations.

In contrast, site-based research programs, such as those found in many national or regionally based museum collections, clinical trials, agricultural research stations, ecological field stations, and national monitoring or disease surveillance programs, typically prefer context-dependent expertise over rigid standardization. In such cases, local experience and personal judgment are trusted and used rather than anonymous global criteria for defining data structures (Chapters 4, 6, 7).

Such site-based studies nonetheless repeat their measurements with the aim of creating context-dependent generalizations. Creating and testing a general causal model is an aim of replication that includes a trade-off between accuracy/realism and precision of inferences drawn from the model.[27] A priori identification of the intent to make a general casual claim bounds and defines the construction and justification of casual inferences. However, results from different experiments or statistical models that are based on a generalized model also can be used to improve both precision and accuracy (see Chapter 15).

There is as yet little agreement about how to reconcile these two approaches: a formal protocol for a single universal standard versus context-dependent accuracy. One major thread weaving throughout this book is our attempt to examine whether reproducible data and the trust they should generate can even exist without formal, enforced standardization. Because replication in the sense of "doing the same thing again" is so fundamental to the practice of science, it is typically implicit in our scientific work. But because replication is implicit, it is rarely questioned, and the action of replication is more often than not contingent on circumstance: different standards for "adequate" replication are heuristically identified and applied for any particular question, method, or environment. Identifying a single, standard method for replication may thus improve data interoperability but simultaneously limit their usage and applicability to additional research questions outside the one that originally determined the single standard.[28] Making explicit the implicit assumptions about and actions surrounding replication is, however, a necessary first step.

Describing the Reproduction Process

Accurate process metadata—the description and documentation of the processes that were used to create one or more data sets from raw data—is essential to ensure that a data set itself can be reconstructed and subsequent analyses repeated. In principle, all of the steps—not only routine and formulaic applications of computer software (e.g., converting square centimeters to square millimeters) but also those involving human judgment—should themselves be described, recorded, and preserved in a format that would enable them to be subsequently retrieved and reproduced (see details in Chapters 13, 14).[29]

The need to record one's working process to enable the reproducibility of a data set is, however, only one aspect of the broader challenge of interoperability. A review of common errors and screening methods also can help to build trust as we increasingly come to rely on information derived using computers and provided by them. Process metadata and replicable data analysis can further help individuals (or machines) who attempt to address large-scale or complex questions that require the aggregation and intersections of data sets and data streams that are derived from multiple and diverse sources, such as localized databases, Big Data projects, and natural history collections. Development of best practices for data and metadata documentation and for collection and capture of contextual information—within metadata standards instead of haphazardly reporting on additional or auxiliary information—is an area of active research.[30]

Retrieving Objects

Within a given field trip or experiment, an individual observation of a particular object, collected at the lab bench or in the field, is an example of an instance or an individual token. In the context of this book, this individual object is a replicate. If only representations of this specimen, such as a written description, physical measurements, or photographs of it are collected and stored, it can be difficult or impossible to recover additional information about this object. In contrast, a specimen stored in, for example, a gene bank, tissue bank, or natural history collection includes not only the subject of the observation itself but also formally structured data (e.g., specimen labels) and (often) ancillary data such as laboratory or field notebooks and journals. The specimen itself connects the formal and informal data to the individual token collected in the lab or field (Chapters 4, 5).

Adequately curated specimens provide future access to a physical object with which to validate data after long time spans have elapsed, as well as to additional, often crucial, information that can be used to characterize its historical context. In short, theories, models, and definitions can change, but a physical object (aka a rigid identifier) remains in the museum for further reference (Chapters 4, 5). [31] This replicate connects the actions of seemingly separate actors (collectors, observers, experimentalists) and their methods (descriptions, observations, manipulations).

Repeating the Analysis

Data analysis should be the easiest to replicate: simply re-run a given statistical or analytical model on the original, new, or combined data set. This simple recipe, however, hides much complexity.[32] When statistical methods were developed, options were few and calculations were done by hand. Point-and-click, menu-driven software has simplified the calculation process but hides details crucial for replication—for example, what formulas and algorithms were used in what version of software or operating system, and in what order. Software that uses scripting requires users to type and save commands, but scripts tend to be customized for individual projects and also require version control. Keeping track of these analytical workflows is surprisingly difficult yet is crucial for ensuring formal replication (Chapters 13, 14). Further, repeating statistical analyses on a given data set is different from re-running an analytical model or a computer simulation.[33]

Reproducing Combined Results: Meta-analysis

Meta-analysis relies on combining results from multiple analyses of different data sets and has long been used in medical research and to aggregate and synthesize results of clinical trials (see examples in Chapters 11 and 12). In recent years, it has been applied increasingly, and with increasing importance, in ecology, evolution, and other areas of biological research.[34] In brief, meta-analysis is the use of formal statistical methods to combine results from distinct case studies to increase the statistical power of a comparative analysis among these studies. Consequently, meta-analysis depends completely on the reliability of existing data sets and their analyses. Meta-analysis can increase both precision and generality of results, yet the decision about whether the results of individual studies are similar enough to be combined in a meta-analysis is essential to the validity of the result.

GENERALIZATION

Although the path from observation to experiment, to replication, to synthesizing results from multiple experiments, and finally to creation of a general theory often is portrayed in scientific textbooks and the popular press as following a linear, progressive path from initial, specific observations to scientific "laws," this is rarely the case.[35] In truth, generalizing by choosing from among competing analytical models and statistical tool-kits often comes at the outset of a study, and the model chosen dictates which questions are asked and how data are collected, evaluated, and analyzed.[36] Just as there are many meanings of "replication," so too there are a variety of meanings and interpretations of "generalization." In particular, two different operational concepts of generalization suggest different approaches to each of the three meanings and the five stages of replication.

The first operational concept of generalization aims for representativeness as an epistemic value: a sample should represent a wider context and a model should be reduced to an idealized, simple, elegant, and universal description of nature.[37] Such an idealized model is meant to represent a deeper (i.e., a real) process that subsumes the multiple diverse, and often conflicting, empirical details.[38] This idea is encapsulated in Plato's attempt "to save the phenomena," which required contemporary scientists to find a single, constant, and elegant astronomical law to cover the confusing back-and-forth movements of the known planets.[39]

In contrast, the second operational concept of generalization values comprehensive accuracy as an epistemic quality. A general model in this sense is a collection of many different case studies, each represented by a single (idealized) model that accurately describes the complete causal structure and causally relevant objects and traits in that case study. Generalization proceeds by assembling several different, albeit not necessarily consistent, models, each of which comprehensively represents the actual causal structure that results in the outcome of the case study rather than requiring a simple universal law or standard workflow.[40] Generalization as representativeness asserts that a scientific representation cannot ignore the common structure underlying the various scales, entities, and dynamics. In contrast, generalization as accurate comprehensiveness asserts that a scientific representation cannot ignore nature's multifaceted, context-dependent complexities, as encapsulated in Darwin's famous paragraph on "endless forms most beautiful."[41]

One would expect both representativeness and comprehensiveness to co-exist in any rigorous scientific generalization. In the same context and for the same question, however, representativeness and comprehensiveness often suggest very different protocols and co-exist uneasily, if at all. Design and execution of research programs, and the perceived importance of and approach to replication, typically differ depending on an individual's implicit adoption of one or the other of these two senses of generalization (for details and examples, see Chapter 15). If the emphasis is on representativeness among experimental or observational replicates and on models that represent universal processes, the starting point for a study is the construction of a formal (often mathematical) representation of the model and a specific (ideally randomized) sample design. Such an emphasis on null hypotheses and null models implies that all assumptions and prior knowledge should be eliminated to the extent possible. Data collection focuses only on those observations and key data necessary to test the null hypothesis (Chapter 9).[42]

On the other hand, if generalization implies a coalescence of case studies, the emphasis will be on actively noticing and describing, in great detail, as many variables and processes as possible for each case study, whether or not they are needed to test a particular hypothesis. Whereas generalizing for universal models uses such case-study data only in the initial stage of a research program, when generalization is based on a comprehensive coalescence of case studies it can be misleading to draw general conclusions from data collected under the assumption of a universal or representative model.[43]

REPLICATION AND AN EPISTEMOLOGY OF GENERALIZATION

The possible gap between these two views is highlighted in spatial replication of field surveys: the actual location at which replicates are placed depends on one's epistemological stance regarding generalization. Generalization as a universal model requires that spatial replicates be defined a priori by a specified statistical design (e.g., a grid of longitude and latitude, as in the case study of marine monitoring described in Chapter 8).

In contrast, generalization from coalescent case studies often places spatial replicates in empirically different and incommensurable points on the ground, based contingently on environmental conditions and biological interactions that define that point locality (e.g., appropriate conditions for vegetation patches in the desert, as discussed in the case study presented in Chapter 7). The gap between these operational concepts of generalization will be set in sharp relief when the sampling locations based on contingent variables are geographically different from locations specified randomly by a formal statistical design.[44] In the former case, a replicated study could be analyzed using the identical framework or model used for the initial study. But in the latter case, new spatial locations and new covariates will require analyses with different methods and models, which have the potential to yield substantially different results and conclusions (for examples of such mismatches, see Chapter 16).[45] In fact, strict repeatability is practically impossible in the latter case.[46] In Chapter 17, we discuss possible experimental designs that may help bridge these gaps.

In the latter stages of replication—moving from raw data to working data sets, creating metadata, repeating quality assurance and quality control procedures, and re-running models—epistemological approaches to generalization engender only a few differences. One is in the use of background information for assessing the reliability of analytical results, and another is in extrapolation from the specific to the general. When generalization is used to generate representative, universal models, a robust theory, independently verified at statistically replicated times and sites, can be used to justify inductive inference from a single or a few populations to most others. In contrast, when generalization proceeds from coalescing case studies, induction is of little use. Rather, interpretation and consensus link a collection of case studies to a single, composite picture of reality. Looking ahead to the case studies that form the core of this book (Chapters 4–14), we find a continuum between these approaches. Thus, we suggest that continual and critical discourse between universalists and collectors of case studies can yield productive new insights.

NOTES

1. Shapin and Schafer (1985: 225).
2. The important distinction between individual objects a researcher encounters and the general classes of objects that researchers theorize about has deep philosophical roots that pervade all of science. The individual chickadee visiting a particular bird feeder is an instance of the more general type "chickadee," which itself is an instance of the type "bird." In fact, the very notion of an instance presupposes the existence of a type (or a generalization); see Chapter 2 for a particularly clear example.
3. Robinson (1987: Fragment 12).
4. Shapin and Schafer (1985: Chapters 9, 10).
5. Schickore (2010).

6. Collins (1992) wrote on the general theoretical challenge of replication, focusing on case studies in physics (cf. Hacking 1983), and two National Research Council reports (NRC 1995, 2003) were dedicated to the challenges of
replicating (in the sense of retrieving, aggregating, comparing, and sharing)
data across the biological sciences. Yet, as we write, there has recently
emerged a more concerned and practical discussion of the lack of replication
in the life sciences. Hayden (2013), writing in *Nature,* claimed that weak
statistical standards are leading to a "plague of non-reproducibility in science."
For more detailed examples, see Russell (2013) on funding policies of science;
Ioannidis (2005) and Ioannidis et al. (2011) on epidemiology; Sanderson
(2013) on reproducing chemical reaction pipelines; Müller et al. (2010) on
long-term ecological research; Vittoz et al. (2010) on monitoring vegetation
change; and Charron-Bost et al. (2010) on bioinformatics.

7. We thank an anonymous reader for this comment.

8. We thank Eva Jablonka for this comment.

9. For example, compare the opening of a well-known ecology textbook ("The
ultimate subject matter of ecology: the distribution and abundance of organisms: *where* organisms occur, how many occur there and *why*" [Begon et al.
2006: 1; italics in original]) to that of the *Fundamentals of Physics* textbook
("What is Physics? Science and engineering are based on measurements and
comparisons. Thus we need rules about how things are measured . . . we
measure each physical quantity in its own units, by comparison with a *standard*" [Halliday et al. 2005: 2; italics in original]). The former emphasizes
particular places and causes, while the latter emphasizes universal rules and
standards. Ellis and Silk (2014) argue for the continuing necessity of replicated experiments in physics lest its validity as a science be undermined by
untestable, albeit elegant and explanatory, theories.

10. Michener and Jones (2012).

11. One could go further and determine whether statistically "significant" results are reproducible with repeated measurements.

12. We wish to thank Sivan Margalit for her original thoughts and meticulous
work, Abraham Meidan for his generosity in sharing his helpful software
(WizWhy), and both for discussing these questions in depth and examining
some of the case studies in this book.

13. When asking whether a study can be replicated, one should examine the aspect of accurately identifying the actual cause that makes a difference in the
world as well as its more generalizable aspect, including its statistical significance (Woodward 2003; Strevens 2004). Replication is unlikely if a causal
process occurs in nature so rarely that it is not biologically meaningful or if
it occurs in statistically significant numbers only because of the well-known
bias in the literature for publishing positive (and unexpected) results (e.g.,
Sterling 1959; Dwan 2008; Franco et al. 2014). On the other hand, replication
is more likely to occur if experiments are designed to identify "real" causal
processes or if the results are actually random (such as when tossing dice).

14. This and the subsequent discussion in this paragraph is based on definitions and etymological information from the *Oxford English Dictionary* online (http://www.oed.com/), accessed 6 November 2013.
15. See Cassey and Blackburn (2006) for a similar discussion focused on ecological research.
16. For additional discussions and specific definitions of "the same" with respect to location, see Chapter 16, as well as Kohler (2012); Shavit and Griesemer (2009, 2011a, 2011b). Helm (2009) and Helm and Visser (2010) provide a parallel discussion with respect to "the same" time or season.
17. We thank Jonathan Shaked for this example.
18. Compare Mayr (1959) with Winsor (2003).
19. Strasser (2008).
20. Cassey and Blackburn (2006).
21. Even repeated analysis of DNA sequences—arrangements of individual objects (base pairs) for which repeated sampling of replicated strands should in theory yield identical results—have variability that depends on mutation rates.
22. Cassey and Blackburn (2006).
23. Of course, repetitions as experimental replicates can be made at the same time, but those would be in a different location (even if differing only by millimeters). We consider even experimental replicates at two ends of a lab bench or on two benches within a single growth chamber or greenhouse to be different "locations." The importance of this has long been recognized by plant biologists, who routinely, during an experiment, randomly reassign pots to different greenhouse benches at regular intervals or move around petri dishes of *Arabadopsis* within growth chambers. When this is not possible, benches, greenhouse modules, or growth chambers are treated as "blocks" during data analyses, similar to fields in which agronomic trials are established. See Gotelli and Ellison (2012) for more discussion of experimental designs to account for block effects.
24. Shavit and Griesemer (2009, 2011a, 2011b); Shavit and Ellison (2013).
25. In this context, "quality control" refers to a set of formal procedures that are applied to raw data or a data stream so that the resulting scientific data set adheres to certain agreed-upon criteria that include quality and epistemic values. For example, correcting typographic mistakes or mismatches between the name of a location that appears on a map and one given by an investigator are examples of quality assessment that control for consistency: one name = one geographical location. "Quality assurance" refers to formal procedures that are intended to ensure that the scientific process (before the work is complete) meets certain requirements and epistemic values. For example, sending two different teams to re-locate or re-sample a previously sampled location helps to reduce bias resulting from an individual's experience or ability.
26. For examples, see Grinnell (1910) and Griesemer and Gerson (1993) on early natural history collections; Campbell et al. (2013) on the National Ecological

Observatory Network (NEON); Strasser (2008) on GenBank; and Leonelli (2013) on genetic collections of plants.

27. On the trade-offs between accuracy, precision, and generality, see Levins (1966, 1968) or Weisberg (2006), and compare them with Orzack and Sober (1993).

28. Interoperability is defined by the IEEE (1990: 42) as "the ability of two or more systems or components to exchange information and to use the information that has been exchanged." In practice, this means that a query to a system will be quickly answered by an automated exchange of information between machines rather than relying on human judgment or labor.

29. Documenting such "scientific workflows" is an active area of research in computer science. See Ellison et al. (2006); Jones et al. (2006); Ludäscher et al. (2006); Boose et al. (2007); Osterweil et al. (2010); Lerner et al. (2011); Zhao et al. (2012).

30. See Michener et al. (1997) and Michener and Jones (2012) for relevant examples.

31. Kripke (1972).

32. Bowker (2005); Humphreys (2007).

33. A large body of important philosophical discussion has focused on models in general, and on models in biology in particular. For examples, see Hesse (1966); Levins (1966); van Fraassen (1980); Lloyd (1988); Griesemer (1990); Wimsatt (2007); Weisberg (2013).

34. Koricheva, Gurevitch and Mengersen (2013).

35. See Shrader-Frechette and McCoy (1993) for further discussion.

36. On modeling and simulation strategies see Humphreys (2007); Weisberg (2007, 2013).

37. See Rheinberger (1997). Weisberg (2007) provides an analysis of model idealization that uses a Galilean idealization strategy, promoting simplicity over completeness.

38. Richard Levins (1966) argued for three main explanatory desiderata—generality, accuracy, and precision—and a trade-off between them for building ecological models. We agree with Weisberg's (2006) translation of Levins's "accuracy" into "realism."

39. Duhem (1969).

40. Following Weisberg's (2007) analysis of model idealization, this second type of generalization employs multiple-model idealization, promoting completeness of partial truths (i.e., "lies") over an intentionally simplified, abstracted, and single Galilean model. Echoes of this approach are found in the work of Levins (1966: 20), who asserted that in the generalizations and models (of evolutionary ecology) "our truth is the intersection of our lies." Ives (2013: 9) similarly asserted that "ecology is a collection of case studies."

41. The oft-cited paragraph reads: "There is grandeur in this view of life, with its several powers, having been originally breathed by the creator into a few forms or into one; and that, whilst this planet has gone cycling on according

to the fixed law of gravity, from so simple a beginning endless forms most beautiful and most wonderful have been, and are being, evolved" (Darwin 1859: 445).

42. Strictly speaking, the "test" is a measure of the probability *given* the null hypothesis. If not, the null hypothesis is rejected. But this does not provide comparably strong (i.e., 95 percent) support for the alternative(s). See chapters in Ellison et al. (2014) for additional discussion.

43. Shrader-Frechette and McCoy (1993: 1); Beever (2006).

44. Shachak et al. (2008).

45. Contrast Loya et al. (2004a, 2004b) and Loya (2007) with Rinkevich (2005a, 2005b).

46. Shavit and Griesemer (2011a).

## CHAPTER TWO

Borges on Replication and Concept Formation

Yemima Ben-Menahem

The prism of three stories by Jorge Luis Borges is used to examine a number of problems pertaining to concept formation. Concepts are generalizations that ignore case-specific details, thus facilitating the perception of similarity, repetition, and replication. Furthermore, the meanings of concepts are highly sensitive to context. These observations raise two pertinent questions: To what extent can concepts be grasped across linguistic, historical, and cultural barriers? To what extent can experience be conceptualized "correctly" so that the categories used "cut nature at its joints"?

General concepts are crucial for science—without them, scientists cannot make sense of natural laws, which subsume numerous instances under general patterns. The very notion of an instance presupposes a generalization, or type (see Chapters 1, 13, 14). Moreover, the notions of recurrence and replication are also premised on the ability to subsume individual cases under general concepts. Barring this ability, there would be no way to learn from experience. Scientists and non-scientists alike need to strike a delicate balance between the level of detail required for an accurate description of an individual case and the level of generality required to yield conclusions about other cases.[1]

Scientists are innovators. They survey unfamiliar territories and invent new languages to describe them. It often happens, however, that a scientist discovers that he or she is not the first to enter a new territory; a mathematician, philosopher, or poet had already been there, exploring the same terrain by different means. Physicists, for example, often discover that their problems have already been solved by mathematicians. Biologists often discover that aspects of their questions have been explored by chemists, climatologists, geologists, or even economists, among many others. And all of us occasionally learn that the problems we are struggling with have already been addressed by writers and poets. Note that the notion of anticipation touches upon the problem of replication. For when comparing ideas expressed by different authors we may wonder whether they are in fact one and the same, merely similar, or perhaps similar only in hindsight; are they replicates, repetitions, or reproductions? In this chapter, I examine three stories by Jorge Luis Borges: "Funes: His Memory," "Averroës's Search," and "Pierre Menard, Author of the Quixote."[2] Each of these highlights the intricate nature of concepts and replication in the broad sense.[3]

The common theme running through these three stories is the word-world relation and the problems this relation generates. In each story, Borges explores one aspect of the process of conceptualization, an endeavor that has engaged philosophers ever since ancient Greece and is still at the center of contemporary philosophy of language and philosophy of mind. Together, Borges's stories present a complex picture of concepts and processes of conceptualization. They all confront the question of what is involved in having a concept and what could undermine the capacity of conceptualization, a capacity that we are fortunate to have and would be hard pressed to do without. According to Borges, conceptualization presupposes replication, but, at the same time, what can be considered to be replication depends on the concepts that are employed. Because scientists conceptualize the world and generalize from instances, Borges's stories have important lessons for how scientists construct ideas and replicate their studies.

"FUNES: HIS MEMORY"

"Funes: His Memory" examines memory and forgetfulness.[4] People admire good memories and aspire to improve them. Loss of memory in the wake of trauma or dementia is thought of as one of the worst things that could happen to us.[5] The ideal, it seems, is perfect memory. Not everyone would agree, however. For example, in praising the ability to forget, Friedrich Nietzsche asserts that whereas memory chains us to our past, forgetfulness is a precondition of fully experiencing the present and fully experiencing life.[6] But Nietzsche points only to the psychological-existential benefits of forgetting, which are only part of the story.[7]

In Borges's story, Funes is blessed with perfect memory. Ironically, it was an accident, the very accident that left him paralyzed, that transformed his memory and brought it to perfection. Borges says the story is a metaphor for insomnia, a condition that he apparently was painfully familiar with, and from which Funes suffers as well. A sleepless person is trying to let go of memories, reflections, worries, and the like, but is unable to do so. Just like Funes, the sleepless one is flooded by memories. These disadvantages of perfect memory are psychological, but Borges does not stop there. Rather, he refocuses his story on the philosophical question of conceptualization.

The peculiarity of Funes is not merely that he remembers more than a "normal" person does or that he forgets less. Between his memory and the memories of everyone else, there is a qualitative difference: whereas we abstract and generalize, Funes remembers events as instances, each in its unique singularity. Each impression is recorded individually, with the full richness of its hue, sound, texture, and time, the emotions it evoked, and so on. "Funes remembered not only every leaf of every tree in every patch of forest, but every time he had perceived or imagined that leaf. . . . His own face in the mirror, his own hands, surprised him every time he saw them."[8]

At this level of detail, no two memories are alike and nothing ever repeats itself. (This is directly comparable with the level of detail presented in the execution trace of a reproduced scientific analysis; see the data derivation graph in Chapter 14). These memories in their radiant individuality cannot be subsumed under general concepts, which, by their very nature, require the omission of detail and distinction. Precisely because people do not have perfect memories, general concepts help us absorb experiences and organize them in economical ways.[9]

But Funes has no need for abstraction and economy; he can handle the richness that ordinary people cannot. Because he has no need for generalities, he can work with individual names. Indeed, he embarks on two monumental projects of naming: giving a name to each image of his memory and giving a name to each number.[10] However, these projects begin to reveal the conceptual drawbacks of Funes's perfect memory. The naming system Funes has invented erases all traces of structure, mathematical or other. Because this naming system has no generic terms (or types), such as "dog," it has no way of indicating the relationships between instances of the same type. Similarly, the naming and numbering system is completely opaque to the intricate relations between different numbers. The real problem, it turns out, is not that, because of his extraordinary memory, Funes is unable to sleep but rather that he is unable to think.

Borges explored the same question—how the complex world of experience could be conceptualized—in his non-fiction as well. He described how the seventeenth-century scholar John Wilkins invented an analytical language: "He divided the universe into forty categories or classes, which were then subdivided into differences, and subdivided in turn into species. To each class he assigned a monosyllable of two letters; to each difference, a consonant; to each species, a vowel. For example, de means element; deb, the first of the elements, fire; deba, a portion of the element of fire, a flame. . . . The words of John Wilkins' analytical language are not dumb and arbitrary symbols; every letter is meaningful, as those of the Holy Scripture were for the Kabbalists."[11]

In these few lines Borges alludes to the fundamental riddle of language: is there a natural way of organizing the world, or is language necessarily arbitrary and conventional? One may distinguish between two aspects of the problem: the correspondence between the symbol and what it symbolizes and the deeper problem of the nature of the categories. Borges touches upon the first of these questions by comparing Wilkins's analytical language to the Kabbalists' belief in the possibility of a correct, non-conventional language whose terms express the essence of things. Borges does this even more explicitly in the beginning of his essay on Wilkins by mocking a lady who declares that "the word *luna* is more (or less) expressive than the word *moon*."[12] But even if one fully accepts the conventionality of language in this sense, there remains the problem of the nature of the categories: are they natural or conventional? It is this latter problem that is of primary concern to Borges.

Borges notes that there is no classification of the universe that is not arbitrary and speculative.[13] Yet, an arbitrary language is still a most powerful tool of representation. Borges illustrates this arbitrariness through the classification of animals in a fictitious encyclopedia: "(a) those that belong to the emperor; (b) embalmed ones; (c) those that are trained; (d) suckling pigs; (e) mermaids; (f) fabulous ones; (g) stray dogs; (h) those that are included in this classification; (i) those that tremble as if they were mad; (j) innumerable ones; (k) those drawn with a very fine camel's-hair brush; (l) etcetera; (m) those that have just broken the flower vase; (n) those that at a distance resemble flies."[14]

Foucault opens his preface to *The Order of Things* with his reaction to this passage: "This book first arose out of a passage in Borges, out of the laughter that shattered, as I read the passage, all the familiar landmarks of my thought—our thought, the thought that bears the stamp of our age and our geography—breaking up all the ordered surfaces and all the planes with which we are accustomed to tame the wild world of existing things, and continuing long afterwards to disturb and threaten with collapse our age-old distinction between the Same and the Other."[15]

Foucault goes along with the destabilizing message of Borges's essay. The original French title of his book—*Les mots et les choses*—(literally: *Words and Things*) reveals much more clearly that his subject is the relation between words and objects, language and reality. By telling the history of various conceptions of the word-world relation, Foucault not only presented alternatives to our own conception of the relation between word and object but also challenged the view that any one of the many possible conceptions of the word-object relation is the correct one.

Scientists also would burst into laughter when reading Borges's fanciful description of the classification of animals but not because they would draw the same conclusion as Foucault. Rather, most scientists adhere to a "natural classification" that reflects the structure of nature.[16] The problem with most of Borges's categories of animals is not that they are empty; there are indeed (or could be) animals belonging to the emperor or animals who have just broken the flower vase. The problem, rather, is that these categories are useless from the scientific point of view—they do not figure in any law of nature and do not underpin predictions and explanations.[17]

In sum, general concepts require repetition—the perception of instance x being of the same kind (or type) as instance y. This perception, in turn, requires abstraction from detail, a capacity that Funes lacks. When we use concepts, and sometimes even invent new ones, we confront the problem of objectivity. The names we give our categories can be arbitrary, but the categories themselves, unlike those of Borges's animals, must not be arbitrary. Scientific vocabularies make sense only if they figure essentially in a lawful representation of our varied experience.

"AVERROËS'S SEARCH"

Borges examines another aspect of the word-world relation in "Averroës's Search."[18] Averroës, the protagonist, is capable of conceptualization but lacks some specific concepts that exist in other languages. Rather than dealing with the general question of what is involved in concept formation, "Averroës's Search" is concerned with cultural and linguistic barriers that stand in the way of both understanding and perception. The story addresses the ability to understand the "other" across borders of time, place, language, and culture. As in "Funes: His Memory," replication and recognition of sameness are at the center of "Averroës's Search."[19]

In this story, Averroës is puzzled by what Aristotle means by the terms "tragedy" and "comedy." Averroës thinks that the solution might be in one of his books, which he picks up out of scholarly habit, but he is unable to concentrate on it because there are noisy children playing underneath his window. Having worked all day, Averroës attends a dinner, where one of the guests, a traveler who claims to have been to China, tells his learned audience about his experience. He describes a crowded house of painted wood with rows of cabinets and balconies and a raised terrace: "The people on this terrace were playing the tambour and the lute—all, that is, save some fifteen or twenty who wore crimson masks and played and sang and conversed among themselves. These masked ones suffered imprisonment, but no one could see the jail: they road upon horses, but the horse was not to be seen; they waged battle, but the swords were of bamboo; they died and they walked again."[20]

The host of the dinner responds to this description by stating that "the acts of madmen are beyond that which a sane man can envision."

The traveler protests that "they were not madmen . . . they were present-ing a story." That is, the people (actors) were showing the story rather than telling it. But the dinner guests neither understood how this could be nor appeared to want to understand it.[21]

Like the other guests, Averroës does not understand how a story could be shown, not told.[22] Furthermore, Averroës does not see any con-nection between the odd story he has just heard and the terms "tragedy" and "comedy" in Aristotle's *Poetics*. Borges ends the story with an explicit assertion of the importance of repetition and replication: "In the preced-ing tale, I have tried to narrate the process of failure, the process of defeat. . . . I recalled Averroës, who, bounded within the circle of Islam, could never know the meaning of the words tragedy and comedy. . . . As I went on, . . . I felt that the work mocked me, foiled me, thwarted me. I felt that Averroës, trying to imagine what a play is without ever having suspected what a theatre is, was no more absurd than I, trying to imag-ine Averroës. . . . I felt . . . that my story was a symbol of the man I had been as I was writing it, and that in order to write that story, I had had to be that man, and that in order to be that man I had had to write that story, and so on, ad infinitum."[23]

Borges hides additional clues in "Averroës." For example, the children who had distracted him by playing outside his study were play-acting, albeit without a stage or the concept of a play. "One of them, standing on the shoulder of another, was clearly playing at being a muezzin: his eyes tightly closed, he was chanting the muezzin's monotonous cry, There is no God but Allah. The boy standing motionless and holding him on his shoulders was the turret from which he sang; another, kneeling, bow-ing low in the dirt, was the congregation of the faithful. The game did not last long—they all wanted to be the muezzin, no one wanted to be the worshippers or the minaret."[24]

Because Averroës cannot conceptualize a theater or a play, he can-not perceive the children's game for what it was. Instead, he sees children playing, quarrelling, chanting, and consequently disturbing him. Al-though technically correct, these descriptions miss the mark. Ironically, Averroës is right when he tells himself that "what we seek is often near at hand," but he does not, and in fact cannot, realize how near it is.[25]

Averroës's lesson pertaining to perception is of fundamental impor-tance. Concepts shape perception no less than perception shapes concepts, and the processes of perception and conceptualization must occur in parallel, not sequentially.[26] The upshot of "Averroës's Search," that concepts play an essential role in perception and understanding, is applicable to scientific inquiry as well.[27]

There are two linked issues involved in "Averroës's Search": the role of concepts and the predicament of cultural barriers. The first of these has attracted a great deal of philosophical attention ever since the classical controversy over the status of universals, whereas the second gained additional attraction in the twentieth century.[28] For example, throughout his philosophical career, Ludwig Wittgenstein struggled with the question of whether there is a preferred conceptual system, that is, one that is more suitable than others for the description of the world. He was inclined to answer this question in the negative, and he invented hypothetical scenarios designed to show that alternative conceptual systems could be as useful as our own. Eventually, he may have come to the conclusion that the very question about the correspondence between language and reality is an empty question, one that transgresses the limits of sense.[29]

In the real world, even though general laws of nature apply everywhere, there are many people and cultures that have languages and customs that differ from one another. How reliable is one group's understanding of another's? Wittgenstein addresses this predicament in a detailed critique of Sir James George Frazer's *The Golden Bough*, the classic comparative study of myth and religion.[30] Whereas Frazer interpreted rituals and myths of distant cultures as scientific theories (false, of course, and created by mistaken, naïve, or irrational "others"), Wittgenstein asserted that rituals and myths are symbolic expressions that are not meant to be literally true or false. Wittgenstein claimed that Frazer had overlooked the fact that his own (nineteenth-century Scottish) culture, too, had numerous rituals (prayer, kissing the pictures of loved ones, etc.) that do not function as scientific procedures. Interpretations such as Frazer's, which turn these practices into errors, are wanting, Wittgenstein maintained, not only from the cognitive but also from the moral, point of view.[31]

Even accepting Wittgenstein's point, however, interpretation remains an extremely difficult task. How does one know whether an expression has a literal or metaphoric meaning for a speaker? Whether an action has a causal or a symbolic role? Whether the associations surrounding a phrase used by one individual in one culture are the same as the associations surrounding one used by another individual in a different culture? Borges touches on these issues in "Pierre Menard, Author of the Quixote."

"PIERRE MENARD, AUTHOR OF THE *QUIXOTE*"

Of course, Miguel de Cervantes was the author of *Don Quixote*. In Borges's story, Pierre Menard set himself the formidable task of re-writing certain parts of that book. What does re-writing mean here—replicating, repeating, or reproducing (copying)? Merely reproducing (copying) the work would not make Menard its author, of course. On the other hand, any changes he would make while replicating or repeating it would make him the author of another work, not the author of the *Quixote*. This is the fundamental question about replication, taken to its extreme. In trying to explain it, Borges only makes the question even more paradoxical: "Pierre Menard did not want to compose another Quixote, which surely is easy enough—he wanted to compose the Quixote. Nor, surely, need one be obliged to note that his goal was never a mechanical transcription of the original; he had no intention of copying it. His admirable ambition was to produce a number of pages which coincided—word by word and line by line—with those of Miguel de Cervantes."[32]

How could Menard accomplish this? His first method was to "learn Spanish, return to Catholicism, fight against the Moor or Turk, forget the history of Europe from 1601 to 1918—be Migual de Cervantes."[33] But this method was quickly dismissed, not because it was impossible but because "of all the impossible ways of bringing it about it was the least interesting."[34] The alternatives were even more perplexing: "Being, somehow, Cervantes, and arriving thereby at the Quixote—that looked to Menard less challenging . . . than continuing to be Pierre Menard and coming to the Quixote through the experience of Pierre Menard."[35] However: "Composing the Quixote in the early seventeenth century was a reasonable, necessary, perhaps, even inevitable undertaking; in the early twentieth, it is virtually impossible. Not for nothing have three hundred years elapsed, freighted with most complex events. Among those events, to mention but one, is the Quixote itself."[36]

The question Borges set out to examine in this surrealist tale is the relation between meaning and context, the fact that identical texts can have radically different meanings. He has the narrator of the story tell us that Menard's text, even though identical with the original, is infinitely richer! This paradoxical claim is then illustrated:

> It is a revelation to compare the Don Quixote of Pierre Menard with that of Miguel de Cervantes. Cervantes, for example, wrote the following (Part I, chapter IX):
>
> ... truth, whose mother is history, rival of time, depository of deeds, witness of the past, exemplar and adviser to the present, and the future's counselor.
>
> This catalog of attributes, written in the seventeenth century, and written by the "ingenious layman" Miguel de Cervantes, is mere rhetorical praise of history. Menard, on the other hand, writes:
>
> ... truth, whose mother is history, rival of time, depository of deeds, witness of the past, exemplar and adviser to the present, and the future's counselor.
>
> History, the mother of truth!—the idea is staggering. Menard, a contemporary of William James, defines history not as delving into reality but as the very fount of reality. Historical truth, for Menard, is not "what happened"; it is what we believe happened. The final phrases—exemplar and adviser to the present, and the future's counselor—are brazenly pragmatic.[37]

At the end of the story, Borges suddenly diverts our attention from the paradoxical possibility of rewriting a text to the very realistic circumstance of reading it.[38] The paradox has perhaps gone but not the irony. For hasn't the new technique discovered by Menard—the technique of "deliberate anachronism and fallacious attribution"—been at the very heart of reading all along? Is our own context, conceptual, historical, intellectual, and linguistic, not integral to our reading and interpretation? No matter how hard we try to bracket this context and read the Quixote or any other work precisely as it was written, we are bound to fail.[39]

ACCURACY, GENERALITY, AND THE CHALLENGES OF REPLICATION

Borges's three stories highlight the intricate nature of concept formation, replication, memory, and understanding. Funes has perfect perception and memory, but, precisely because of this gift, he is unable to understand general concepts. Furthermore, concepts require context, both cultural and temporal. Averroës, who has no knowledge of theater, cannot help but misinterpret Aristotle's notions of tragedy and comedy. Pierre Menard's *Quixote*, written in the twentieth century, means something completely different from that of Cervantes, even though the words are identical. And in his non-fiction, Borges challenged the seventeenth-century attempt to discover an ideal language whose predicates would match natural kinds and natural categories, and would thus cut nature at its joints. The other chapters in this book illustrate in great detail the significance and importance of contextual factors in the sciences.

With Funes, Borges identifies the need to strike a delicate balance between the level of detail required for an accurate description of an individual case and the level of generality required to yield conclusions about other cases and types (see also Chapters 13, 14). Moreover, the notions of recurrence and replication (e.g., of an observation or an experiment) are premised on our ability to subsume singular cases under general concepts (see Chapters 1, 4–10). Barring this ability, there would be no way of learning from experience (see especially Chapter 15). General concepts are thus crucial for science. Through Averroës, John Wilkins, and Pierre Menard, Borges reminds us that our concepts not only describe and organize our experiences but also actually shape the world for us. Our concepts are ever changing; so, too, are our perceptions. A universal conception of replication, like that of a universal language that carves nature at its joints and ensures perfect communication between all human beings at all times, is perhaps a noble ideal, but, for better or worse, it is one that may be impossible to realize.

NOTES

1. Levins (1966, 1968) provides a parallel example of the trade-off between generality and precision.
2. Jorge Luis Borges (1899–1986), the great Argentinian writer and poet, has been described as the most important Spanish-language writer since Cervantes. He lived most of his life in Buenos Aires, the site of many of his stories, where he also served as director of the Argentine National Library. His remarkable erudition in history, philosophy, literature, religion, mysticism, and more is reflected both in his short stories and in the numerous essays he wrote. His distinctly ironic style in both of these genres sometimes makes it difficult to identify the fictitious aspects of his non-fiction or the true-to-life basis of his fiction. His writings investigate deep philosophical

problems such as the concepts of infinity and time and have therefore been acclaimed not only as superb literature but also as inspiring for the philosophically minded reader.

3. All three of these storied first appeared in the 1940s and were translated by Andrew Hurley in Borges (1998). Page numbers in subsequent quotations from these stories refer to that edition.

4. "Funes: His Memory" is the fictionalized story of Borges's meetings in Uruguay with Ireneo Funes. Funes, as the result of a head injury sustained after falling from his horse, gains the ability to remember everything possible.

5. For example, Søren Kierkegaard (1843: 19) wrote: "When one lacks the categories of recollection and repetition, all of life dissolves into an empty, meaningless noise."

6. "He who cannot sink down on the threshold of the moment and forget all the past, who cannot stand balanced like a goddess of victory without growing dizzy and afraid, will never know what happiness is" (Nietzsche 1874: 62).

7. Nietzsche recommends a balance between two indispensable orientations: "The unhistorical and the historical are necessary in equal measure for the health of an individual, of a people and of a culture" (Nietzsche 1874: 63).

8. Borges (1998: 136).

9. Quiroga (2012) provides a comparable example from neurological research into concept formation.

10. "He would say, for instance, 'Máximo Pérez,' instead of seven thousand and fourteen (7014)" (Borges 1998: 136).

11. "John Wilkins' Analytical Language," in Borges (1999: 229–232).

12. Borges (1999: 229).

13. Ibid. 231.

14. A Chinese encyclopedia that Borges named *The Heavenly Emporium of Benevolent Knowledge* (Borges 1999: 231).

15. Foucault (1970: xv).

16. That is, the organization of the "real world" is not based on arbitrary categories. Rather, normal science aims to produce what Pierre Duhem (1906) called a natural classification. Some philosophers call this a classification that carves nature at its joints. The metaphor, but not the exact phrase, is traced to Plato's *Phaedrus* 265–266.

17. This problem is illustrated by the standard deductive-nomological model of explanation. In this model, the sentence describing the event for which we wish to provide an adequate explanation is entailed by the sentences expressing the laws governing this event and its initial conditions. Clearly, for the derivation to go through, the predicates appearing in the description of the explained event must match those appearing in the explanatory laws and conditions. If we fail to describe the event in the appropriate language of the laws and use some alternative description instead—even a correct alternative description—there will be no deduction, and consequently, no explana-

tion or prediction. Scientific explanation is sensitive to the descriptions we use and to the categories these descriptions refer to (e.g., Davidson 1967). This sensitivity does not make science subjective or arbitrary. Rather, by limiting the descriptions we can use to those that refer to categories whose members are characterized by specific lawful behavior, it reflects an objective characteristic of our classifications. At the same time, description sensitivity highlights the centrality of language in the scientific endeavor.

18. "Averroës's Search" is a fictionalized account of some of the challenges faced by the Islamic philosopher Ibn Rushd (1126–1198) as he translates Aristotle's *Poetics*. In particular, Averroës has no familiarity with theatrical plays, as they were not part of his own culture.

19. For an analogous approach applied to scientific modeling, see Weisberg (2013).

20. Borges (1998: 238–239).

21. "No one understood, no one seemed to want to understand" (Borges 1998: 239).

22. The host of the dinner party also asks the traveler whether the people on the Chinese terrace were speaking (in addition to showing). When the traveler answers that "they spoke and sang and gave long boring speeches," the host remarks, "In that case, there was no need for twenty persons. A single speaker could tell anything, no matter how complex it might be" (Borges 1998: 239). The other guests completely agreed and turned to other, seemingly unrelated, subjects. In fact, however, Borges continues to weave threads of replication and understanding throughout "Averroës." For example, the guests discuss the question of metaphor (Does it lose its force when repeated over the centuries?); sin and error ("Only the man who has already committed a crime and repented it is incapable of that crime"); and differences of opinion ("To be free of an erroneous opinion . . . one must at some time have professed it") (Borges 1998: 239–240).

23. Borges (1998: 241).

24. Ibid. 236.

25. Ibid.

26. This lesson is both Kantian and Quinean: "We do not learn first what to talk about and then what to say about it" (Quine 1960: 16).

27. In science, concepts are more useful when embedded in a system of (natural) laws. The discovery of the concept of energy provides a striking confirmation of this insight. Physicists did not first identify energy as a physical entity but then learned that it was conserved (Elkana 1974). Even von Helmholtz (1847), when defining the law of the conservation of energy, did not yet have the concept of energy (the title of von Helmoltz's paper actually is "On the Conservation of Force"). We owe the subsequent understanding of the term (and concept of) "energy" to the discovery of the law, not vice versa. As the scope of the law of the conservation of energy has broadened, the concept of energy also has been redefined. Similarly, as biological studies are

replicated, whether identically or not (see Chapters 4–10, 15, 16), concepts and laws may be expanded or redefined.

28. The problem of universals pertains to the status of general abstract terms such as "beauty," "courage," "health," and "justice" and to the status of properties—what is common to all red things, all tables, all flowers, and so on. Both Plato and Aristotle struggled with this problem throughout their philosophical careers. Despite the significant differences between their positions, both of them are considered to have been realists; that is, they accepted the reality of general concepts as objective and mind-independent entities that would exist even if there were no minds to conceive them. The controversy was further developed and refined in the Middle Ages, when it became customary to distinguish between realists, conceptualists, and nominalists. Like realists, conceptualists see universals as objective, but unlike realists, they consider them to be mental creations. Nominalists deny realism altogether. They maintain that only individuals exist and that the ties between individuals sharing a property, for instance, being red, are only linguistic ties. In their view, there is no objective basis to the generalizations and concepts created by human beings.

29. Wittgenstein's hesitation is manifest in the following cautious formulation: "I am not saying: if such and such facts of nature were different people would have different concepts. . . . But: if anyone believes that certain concepts are absolutely the correct ones, and that having different ones would mean not realizing something that we realize—then let him imagine certain very general facts of nature to be different from what we are used to, and the formation of concepts different from the usual ones will become intelligible to him" (Wittgenstein 1953: II, xii).

30. Wittgenstein (1979).

31. The point of view of Wittgenstein (1979).

32. Borges (1998: 91).

33. Ibid. 91.

34. Ibid. 91.

35. Ibid. 91.

36. Ibid. 93.

37. Ibid. 94.

38. "Menard has (perhaps unwittingly) enriched the slow and rudimentary art of reading by means of a new technique—the technique of deliberate anachronism and fallacious attribution" (Borges 1998: 95).

39. Or, to paraphrase Thomas Wolfe, we just cannot go home again. Here I replicate Shavit (2014), who used this quote as the title of a paper on replication in biological surveys.

## CHAPTER THREE

The Historical Emergence of Replication

REIFYING GEOGRAPHY THROUGH REPEATED SURVEYS

Haim Goren

*The accuracy and precision of maps increase with an area's
geopolitical and scientific importance. The history of repeated site
visits and measurements followed by evolutionary re-creation of site
maps results in a traceable intellectual map of scientific progress
and the accompanying abandonment of an exact copy.*

Replication in its broadest sense is an integral and essential part of scientific research (see Chapter 1), but "replication" has universal application and applicability (see Chapter 2). In this chapter, the importance of
replication is explored for a crucial historical turning point, when new
and progressive scientific measurements of physical locations were being developed. Revisiting a location is of necessary and critical importance when replicating research in the lab or the field, but identifying a
precise location can be surprisingly problematic (see also Chapters 4–8,
15, 16).[1] Geography includes the study and identification of where objects
are located and how they are arranged in space. Whether identifying
spreads of emergent diseases or distribution of genetically distinct populations, we use maps and topographic contours. The maps we use today
are the result of over a millennium of repeated field work, analysis, and
interpretation that provides additional insight into the process of replication. In this chapter, this process of geographic replication and its criteria

of success are illustrated with two examples: the repeated mapping of the city of Jerusalem and the attempt to measure accurately the elevation of the Dead Sea relative to sea level. These examples also reveal multiple motives for repeated exploration and study. Mapping these sites had both historical and religious importance that built on and transcended the geographical data.

BACKGROUND

Until the beginning of the nineteenth century, the Holy Land was virtually a terra incognita from a scientific point of view.[2] Between the end of the eighteenth century, following its "re-discovery" by geographers and scientists in Western Europe, and the beginning of World War I, the geography of and foundations for the scientific study of the region were firmly established.[3]

The approach to describing the geography of the Holy Land was similar in most respects to the scientific study and geographic descriptions by Western scientists of other regions, such as Africa or Latin America, but different in one important respect. In all cases, progress in the mapping of different regions was directly related to their scientific, economic, and political importance to Europe, as well as to their accessibility and the availability of pre-existing information on them.[4] Geographic study of all of these areas was done in the field, involved mapping and other fieldwork, and was conducted in foreign (i.e., non-European) spaces. Results, including new maps and analyses, were published in European countries.[5]

The unique feature in the historical development and geographic conceptualization of Palestine has been its religious significance. Centuries of European biblical scholarship had resulted in the accumulation of a wealth of Christian-inspired interest in the Holy Land. Biblical narratives were a leading motivation for its study, which influenced almost everyone who contributed to clarifying its geography.[6] Consequently, mapping the Holy Land not only was concerned with cartography but also often focused on re-establishing the geographical and historical reality of events described in the Bible.[7]

REPLICATION WHILE MAPPING THE HOLY LAND

In the context of most geographical surveys, new maps improve on older ones by increasing precision and re-producing accurate features while correcting those that were (or have become) inaccurate. However, people are constantly modifying their own environments and altering their geography.[8] Trying to make information about locations (i.e., spatial data) more accurate and precise as the geography is continually being modified suggests that the broad concept of replication may be inapplicable to historical geography.[9]

This dilemma was not part of pre-nineteenth-century geographical studies of the Holy Land because the goal of this earlier work was not to improve the maps but rather to re-discover or copy earlier ones in order to assert the validity of biblical events. The new studies, therefore, represent an important turning point in the history of scientific replication: from identically copying earlier maps, measurements, or experiments to improving them and learning from those changes (see also Chapters 1 and 15).

For over 1,000 years, the Holy Land was discussed, visited, and studied repeatedly, and virtually identical results were published. Greek and Roman historians first attempted to identify, or re-identify, sites mentioned in Scripture. The subsequent "pilgrims and travelers literature" that dates to the beginning of the fourth century continues this tradition of repeated visits to the same places. These "beaten tracks" were regular routes guiding visitors that were established by the Franciscans, who in the fourteenth century established the Custodia Terrae Sanctae.[10] Hundreds of thousands of visitors re-traced these beaten tracks from Jaffa to Jerusalem, to Bethlehem, and to the Dead Sea. Smaller numbers added Hebron or the Sinai Peninsula to their itineraries or traveled from Acre to Nazareth and Damascus. Thousands of their works are known, and for many periods and regions these are the only existing sources for re-constructing the natural and cultural geography of the region.[11]

Yet these early sources provide little actual geographical data or reliable information about the re-visited sites. Yes, the visitors testified to a repeated visit; however, their repetition resembles the efforts of Borges's Pierre Menard (see Chapter 2) in the sense of copying texts that leave us with geographical data almost identical to those provided by earlier informants instead of adding or accumulating any new information or cartographical description. As there were barely any increases or improvements in the accumulated data over more than 1,000 years, the medieval maps, even those pretending to be realistic, represent simply cartographic genealogy rather than cartographical data analysis or actual use of cartographic depictions. Not only can these maps be grouped into historical or contemporary ones; they can also be divided into imaginary and realistic depictions.[12]

These methods of repetitive mapping changed dramatically during the nineteenth century, as exemplified by developments in mapping Jerusalem and measuring the level of the Dead Sea. Thenceforth, the goals of repeated surveys were to improve the maps and obtain better data through repeated surveys rather than to re-identify or re-discover these areas again and again. It is important to note that the change was in the methods and techniques used; the geographical scientific questions and research objectives remained the same.

Mapping Jerusalem

The new geographic studies of Jerusalem began with the production of maps based on two different methods, compilation and geodetics, applied separately or together.[13] Compilation maps were based upon all available material from the various accessible sources, including earlier maps, itineraries, sketches, and limited field measurements. Because earlier maps suffered from inaccuracy and imprecision, most compilation maps of Palestine relied on other sources of information.[14]

Rapid advances in geodetics and the application of triangulation resulted in more accurate transformations of three-dimensional data into two-dimensional maps.[15] The first map of Jerusalem that partially depicts its geographic, natural, and cultural landscapes and was produced from actual triangulated data dates to 1818; the final survey in the nineteenth century was in 1865.[16]

Franz Wilhelm Sieber (1789–1844) created the first quantitatively based map of Jerusalem founded on six weeks of studying the city in 1818 and measuring approximately 200 points (fig. 3.1).[17] This map was praised for its relative accuracy as well as for being a turning point in the cartography of the city.[18] The architect Frederick Catherwood (1799–1854) arrived in Jerusalem in March 1833 but left it in haste in November of that same year. Following his visit, he produced in London a panorama of contemporary Jerusalem and a map of the city. He improved upon Sieber's map; his most notable achievement was to have been the first

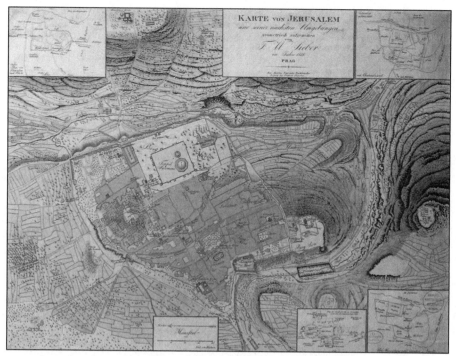

FIG. 3.1. Sieber's Map of Jerusalem (Sieber 1818). Image from GLAM National Library of Israel, and reprinted under the terms of a CC-BY-SA-3.0 license.

European to study and measure the Temple Mount from the inside, including some of its subterranean structures.[19]

Repeated surveys followed by more accurate compilations followed rapidly thereafter. Surveys in 1838 by Edward Robinson and Eli Smith yielded data that were included in a comprehensive compilation map at a scale of 1:100,000 by the Berlin cartographer Heinrich Kiepert (fig. 3.2).[20] A group of British army officers, primarily trained surveyors belonging to the Royal Engineers, accompanied the troops dispatched to Syria in the summer of 1840 as part of the effort to re-instate the Ottoman regime.[21] They measured and mapped the fortifications of several coastal cities, and after a visit to Jerusalem in the winter of 1841, they drafted a map of the city on a scale of 1:4,800.[22] This map was used as the basis for subsequent maps of the city through the mid-1860s.[23]

The final map produced during this historical period was the Ordnance Survey of Jerusalem's 1:2,500-scale map of the city (1:10,000 of the surrounding environs).[24] The primary goal of this 1864 mapping

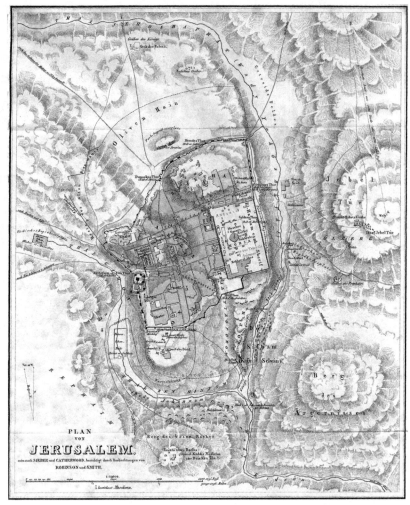

FIG. 3.2. Kiepert's Plan of Jerusalem (Kiepert 1845). Image from Leibniz Institute for Regional Geography, Central Geographical Library, 1979 B 1061-2, and used with permission.

expedition was to improve the city's water supply. For the first time, a map of Jerusalem included a full list of the names of all streets, alleys, squares, and other features. It also devoted special attention to relative elevations of the city and its surroundings, using isohypse lines for the first time in Jerusalem cartography.[25]

The van de Velde "Plan of Jerusalem," first published in 1858 (fig. 3.3), brought to a close a half-century of repeated surveys and re-surveys of

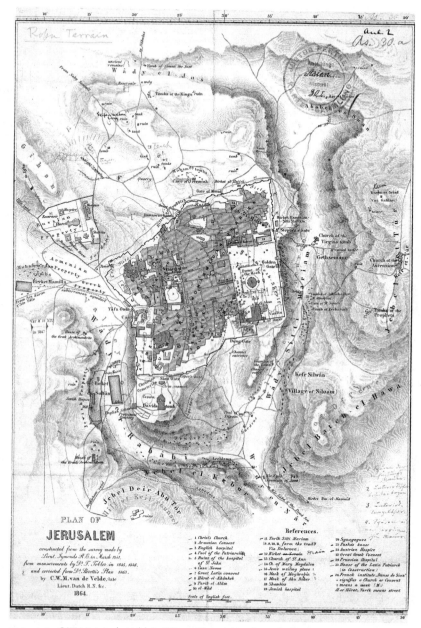

FIG. 3.3. C.W.M. van de Velde's "Plan of Jerusalem from the Survey Map by Lieutenant Symonds R. E. in March 1841" (Gotha: Perthes 1864) (van de Velde 1858). Image from Universität Erfurt, Forschungsbibliothek Gotha (FBG SPK 547_112553222), and used with permission.

Jerusalem that yielded a detailed and accurate map of the city that was used for the coming decades. This map was unique in that it did not use any of the material or data used in the compilation of previous maps. Exact copying was no longer considered valuable for a repeated scientific expedition. At the same time, the new map can be considered either a repeated sample or a new baseline (see also Chapters 6–8 and 15) for future re-surveys of Jerusalem. In that sense, van de Valde's map reveals a historical turning point.

Measuring the Lowest Point on Earth

As with mapping Jerusalem, the initial focus of exploration of the Dead Sea derived from scriptural descriptions of it and from curious and unexplained phenomena attached to it. For example, one main objective was to look for lost cities and existing signs of biblical catastrophes. Scientific explorations of the Dead Sea also were undertaken contemporaneously with the mapping of Jerusalem.[26]

Each re-survey of the Dead Sea included travel, field measurements, and mathematical calculations, and, again as with the mapping of Jerusalem, each involved a different degree of replication and innovation. For example, reaching the Dead Sea by sailing the Jordan River was first attempted in 1835.[27] Two years later, however, two teams, one of which also used a sailboat, measured the level of the Dead Sea[28] and determined that it was substantially below sea level.[29] Confirming this finding through repeated measurements and replication and determining the precise level of the Dead Sea (fig. 3.4) occupied geographers for another 15 years.[30]

Despite the repeated re-surveys, the methods used varied widely. The higher-than-expected temperature of boiling water at the Dead Sea first suggested that the body of water was below sea level.[31] Measurements in the subsequent five years were made using barometers, with variable results.[32] In 1841–1842, a re-survey of coastal cities and fortifications of Palestine was done using triangulation, with the preceding year's Ordnance Survey map of Jerusalem as the starting point. By triangulating lines from Acre to the Sea of Galilee via Safed and from Jaffa to the Dead Sea via Jerusalem, the level of the Dead Sea was estimated at 426.4 m below sea level (fig. 3.5).[33]

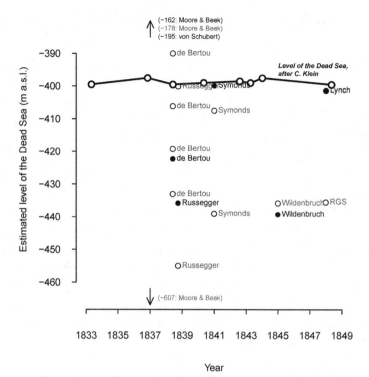

FIG. 3.4. Measurements of the surface level (meters below sea level) of the Dead Sea (1837–1848), compared with current estimates by Klein. Labels in gray indicate approximate values from ranges or estimates provided by the cited authors, whereas labels in black indicate exact values as given by the cited authors. Data compiled by the author; figure drafted by Tamar Sofer and re-drawn by Aaron Ellison.

Two additional re-surveys were completed in the late 1840s. An 1845 topographic profile from Jaffa through Jerusalem to the Dead Sea identified the level of the Dead Sea as 439.27 m below sea level.[34] An 1848 expedition estimated the level at 1,234.59 feet (401.24 m) based on multiple barometric measurements.[35]

CAN PHYSICAL OR HUMAN GEOGRAPHY EVER REVEAL EXACT LOCATIONS?

Accurate mapping—of sea level, city monuments, or the spread of yellow fever, among many other examples—requires accurate locations. If "location" is simply positional information on a particular oblate spheroid

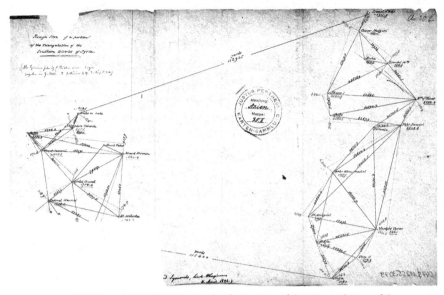

FIG. 3.5. J.F.A. Symonds's "Rough Sketch of a Portion of the Triangulation of the Southern District of Syria" (1842). Image from Universität Erfurt, Forschungsbibliothek Gotha (FBG SPK 547_112553079), and used with permission.

orbiting the sun, accurate and increasingly precise locations should be easy to determine using global positioning system. However, case studies of mapping have revealed that obtaining accurate locational information with repeated surveys is quite difficult and yields different and often incompatible results with each re-survey.

Some of these differences certainly arise from the way the questions are posed and from the cultural or religious perspectives that the investigators bring to their investigations. But others are independent of the researchers' state of mind. Cities are constantly being re-developed and re-made during the intervals between surveys. And, in the case of the Dead Sea, its measured level has changed through time, declining by more than 10 m in the last 15 years of the twentieth century alone.[36] Similarly, the measured production of poultry populations changed due to different vaccination standards at locations upholding different regulations (see Chapter 10), and the presence of different cholera serogroups was detected in human populations using different antibiotics (see Chapter 12). Contemporary re-surveyors and experimentalists probably assume that better equipment leads to more precise measurements, but such precision does not necessarily translate into more accurate measure-ments or greater ability to re-visit precise localities, as the re-measurements of Jerusalem and the Dead Sea have clearly illustrated.

NOTES

1. From the Latin stem *locāre,* meaning to place or to hire, from which we derive "locus," "locate," and "location."
2. Understood to be, variously, Western Palestine, Judea, the geographic area where Jesus Christ lived and died, and the modern-day state of Israel.
3. Ben-Arieh (1972) described the "rediscovery" of the Holy Land in the nine-teenth century. Ben-Arieh (1979, 1987) and Goren (2010) discuss in more detail how European scholars explored the region between 1799—when the French Army under Napoleon invaded Palestine but failed in its cam-paign to capture the Port of Acre—and the beginning of the First World War in 1914.
4. Edney (1997).
5. This is part of the classical model of the development of Western science (Basalla 1967).
6. Goren (1998).
7. For example, scholars went to the Dead Sea not only to map it and determine its elevation relative to sea level but also to re-locate the sunken cities and explain the catastrophe described in the Book of Genesis. Geographers study-ing Jerusalem were attempting to identify the exact locations of buildings and monuments to authenticate sacred events.
8. In the sense of the physical features of the environment.
9. Davies (1968).
10. The Custody of the Holy Land. See Lemmens (1916); Medebielle (1963).
11. Tobler (1867); Röhricht (1890).
12. For examples, see Rubin (1989, 1990, 1999).
13. Geodetics is the branch of mathematics that deals with measuring parts of the Earth or the entire Earth. It accounts for curvature, as well as for spa-tially varying gravitational strength and plate tectonics (the latter of which were unknown to nineteenth-century cartographers). General reviews of sci-entific cartography of nineteenth-century Palestine include Fischer (1939, 1940); Schattner (1951); Tishbi (2001); Goren (2003). More comprehensive bibliographies of the field at the time include Ritter (1850); Tobler (1867). Summaries of the Palestine Exploration Fund, which supported much of this early work, are in Watson (1915); Ben-Arieh (1979); Silberman (1982); Moscrop (2000).
14. Goren (2002) reviews these nineteenth-century compilations. By way of ex-ample, Jean Baptiste Bourguignon d'Anville (1697–1782) excelled in the sys-tem of comparing data and carefully analyzing it, accompanying his maps with memoirs. Still, his maps both of Jerusalem and of Palestine (ca. 1771)

show how poor, unreliable, and coincidental were his sources. See also Godlewska (1999); Bartlett (2008).

15. Edney (1997); Godlewska (1999).

16. Ben-Arieh (1973); Rubin (1999).

17. Sieber (1818).

18. Tobler (1857); Goren (2002).

19. In between Sieber's and Catherwood's maps, other maps were produced
that have been referred to since then only rarely (Westphal 1825; Berghaus
1835; see summary in Goren and Schelhaas 2014). For details of Catherwood's map, see Ben-Arieh (1974). Catherwood is perhaps more famous
for his work exploring the remnants of the Mayan civilization (von Hagen
1968).

20. Edward Robinson (1794–1863) was an American minister and scholar and is
thought of as the father of Holy Land research (Goren 2003). Kiepert's map
(Kiepert 1841) included all information from all previous maps, along with
data collected by Robinson and his traveling companion, missionary Eli
Smith from the American Board of Commissioners of Foreign Missions
(Robinson and Smith 1841). Kiepert himself dedicated much of his life to advancing the mapping of Palestine (Goren 1999).

21. Jones (1973); Goren (2005).

22. Aldrich and Symonds (1841). The map of Jerusalem was later published as
an appendix to the second edition of Williams (1849).

23. For example, Kiepert (see note 20, above) used the Royal Engineers map in a
map of Jerusalem that he published in 1845 based on new measurements
and other data collected by Ernst Gustav Schultz (1811–1852). Schultz served
from 1842 until his death as the first Prussian consul to the city. The map
was published as an addendum to a lecture Schultz presented before the
Geographical Society of Berlin (Kiepert 1845). It is noteworthy for depicting
not only the contemporary city but also historical-geographical data from
scriptural periods (Goren 2003).

   This repetitive and iterative process continued intensively. Following
his visit to the holy city in 1845, the Swiss physician Titus Tobler (1806–1877)
published a map based mainly on the works of Catherwood, Robinson, Kiepert, and the Royal Engineers, to which he added the web of alleys and cul-de-
sacs within the walls of Jerusalem (Tobler 1849). As Tobler was not a trained
surveyor, his sketches and the published map suffer from some inaccuracies
(Goren 1994–1995). Tobler's works, in turn, were used as the basis for the
next map of the city (van de Velde 1858). Van de Velde made sure to use all
existing materials for his maps, including the map of the whole country as
well as the map of Jerusalem and its environs. He traveled to London, Paris,
and other locations before moving on to the cartographic center in the
Thuringian city of Gotha. There he produced his maps at the leading geographical Institute of Justus Perthes Publishers under the auspices of the
German cartographer August Heinrich Petermann (1822–1878). Tobler

added to the map a detailed memoir. Finally, the Sardinian officer and engineer Ermette Pierotti (b. 1821) worked in Jerusalem between 1854 and 1861 and published two maps of the city, to which he added many previously unrecorded details (Pierotti 1860).

24. Wilson (1865) as part of the Ordnance Survey of Jerusalem.

25. Isohypse lines connect points of equal height and barometric pressure, unlike contour lines, which only connect points of equal height. Correcting for local barometric pressures would be particularly important for controlled delivery of water.

26. Among others, the cities of Sodom and Gomorrah are thought to have been on the southeastern shore of the Dead Sea. Some scientific explorations of the Dead Sea were undertaken in the eighteenth century, but significant research began in earnest in the early nineteenth century. The first European since time of the Crusades to circumnavigate the lake was Ulrich Jasper Seetzen (1767–1811). In addition to exploring some of the mysterious phenomena associated with the Dead Sea, he described the life in and around it and also sketched a map of the entire Jordan Valley from the sources of the river to the southern end of the Dead Sea. His map (Seetzen 1810) was based on field observations but not quantitative measurements (Kruse 1854; Schienerl 2000; Goren 2003).

27. This was a non-trivial undertaking that claimed many lives. The first victim was Christopher Costigan (1810–1835), a young Irish theology student. In the summer of 1835, he brought a small boat to Tiberias and launched it on the Sea of Galilee, planning to sail along the Jordan and into the Dead Sea. Although he accomplished his goal, he died in Jerusalem only a few days after having reached the Dead Sea, and he left no evidence or documents of his voyage. His story was known to contemporary travelers, principally through a book of John Lloyd Stephens (1805–1852), who, while visiting Palestine in 1836, re-traced Costigan's route. Another description was provided by the daughter of John Nicolayson (1803–1856), a Schleswig-Holstein-born Protestant missionary employed in Jerusalem by the London Society for Promoting Christianity amongst the Jews. Nicolayson himself had transported the ailing Costigan from the shore of the Dead Sea at Jericho to Jerusalem and related Costigan's story to his daughter, who, several years later, presented it at an 1850 meeting of the Jerusalem Literary Society (Goren 1997).

28. By "level," geographers mean the elevation of the surface of the Dead Sea relative to sea level (defined as zero). Since it was generally accepted that all waters flow downhill to the sea, the idea that a body of water could be below sea level was quite revolutionary.

29. The first team was made up of George Henry Moore and William G. Beek (Moore and Beek 1837; Moore 1913; Goren 2011). The second was a team led by Gotthilf Heinrich von Schubert (1780–1860), who was on a pilgrimage-cum-research voyage with his wife, two students, and a painter (von Schubert 1839; Goren 2011).

30. A complete history of the determination of the level of the Dead Sea is given by Klein (1986). See also Klein (1982); Bookman et al. (2006).

31. Moore and Beek (1837: 456) wrote: "From several observations upon the temperature of boiling water, it [the Dead Sea] appears to be considerably lower than the ocean." The estimate of the temperature of boiling water that they observed,—213°F, made repetition of this study possible.

32. For example, von Schubert (1839) used a barometer and measured the elevation of the Dead Sea to be 195.5 meters below sea level. De Bertou (1839) claimed that the difference in level between the Red Sea (presumably at sea level) and the Dead Sea was 422.27 m, but this was amended in 1842 to 406.13 m (Hamilton 1842). Other measurements summarized by Hamilton (1842) include those of Russegger (1,400 feet = 426.72 m), Wilkie (1,200 feet = 365.76 m), and Symonds (1,311.9 feet = 399.87 m). Additional confusion was injected by Ritter (1850), who referred to de Bertou's measurement in Parisian feet (1,290 Parisian feet = 419.25 m). Some of these differences undoubtedly resulted from different types of barometers, calibrations, and actual locations of measurement.

33. This triangulation was done in 1841 by Lieutenant John Frederick Anthony Symonds (d. 1852) of the Royal Engineers. His measurement was reported by Hamilton (1842); the survey is summarized by Jones (1973) and Goren (2005).

34. As measured by A.A.H. Louis von Wildenbruch, Prussian General Consul to Syria and Palestine from 1843–1848, and reported in Ritter (1846).

35. Lynch (1849, 1852). A contemporary re-telling of this expedition that blends the historical with the religious is Jampoler (2005).

36. Bookman et al. (2014). To add further confusion, a Google search on "level of the Dead Sea" returns values ranging from 1,300 feet (396 m) below sea level (in the *Encyclopedia Britannica* online) to 423 m below sea level on a new map of Israel.

PART TWO REPLICATION IN BIOLOGY
OVERVIEWS AND CASE STUDIES

The Value of Natural History Collections

## CHAPTER FOUR

Natural History Collections as Dynamic Research Archives

Tamar Dayan and Bella Galil

Natural history specimens with detailed locality, environmental, and survey data improve the ability to replicate ecological studies at a wide range of spatial and temporal scales. Results from such studies address fundamental questions in biodiversity and improve its conservation and management.

Natural history collections are archives of biodiversity, snapshots that provide a way to physically retrieve an individual specimen and through it track changes in populations and species across repeatable surveys in time and space. With histories spanning hundreds of years and encompassing huge numbers of specimens, natural history collections are amazing resources for scientific research.[1]

Growing international awareness of the potential effects on humanity due to the loss of biodiversity and the ensuing erosion of ecosystem services has reinforced the value of natural history collections, museums, and herbaria worldwide.[2] Because museums contain records of the past and its dynamics, they increasingly are being used for research into patterns of global biodiversity and global change.[3] After several decades of stagnation, natural history collections-based research has enjoyed rapid growth in recent years as it has entered the mainstream of research in ecology and conservation biology.[4]

The increased interest in natural history collections research also reflects the development and impact of "macroecology." Macroecology takes a broad-brush approach to ecological patterns by examining data collected at continental to global geographical scales in an attempt to reveal dominant patterns and trends in biodiversity.[5] Studies of the projected ecological effects of climatic change on the distribution of species provide an example of how macroecological studies use natural history collections (see examples in Chapter 5). Such studies start by describing changes in the broad distribution of one or more species in the past century or centuries in response to known climatic changes. The results of such studies are then used as inputs to models that forecast changes in the future distribution of the same species as the climate continues to change.[6]

The detailed metadata associated with natural history specimens, including information on when and where they were collected, along with their permanence in collections, contribute to their value in allowing for repetition or replication of historical studies. In this chapter, we summarize the strengths and weaknesses of natural history collections for repeated surveys and other historical studies that require replication. Through a case study of the historical surveys and resurveys of the taxonomic exploration of the marine biota of the eastern Mediterranean Sea, we highlight the relevance of collections for ecology and conservation. Finally, we discuss prospects for future uses of natural history collections in the context of replicated research.

NATURAL HISTORY COLLECTIONS AS THE PRIMARY LONG-TERM RECORDS OF BIODIVERSITY

Specimens in natural history collections testify to the presence of species at given localities and specific times. If the locality data associated with particular specimens are adequately specified (see Chapters 5, 15), there likely is sufficient information to specify locations for subsequent resurveys.[7] The presence of a retrievable physical object—the unique specimen known to have been present at a specific locality—allows subsequent researchers to ask whether the same species is still found in the same place or whether it has been replaced by different ones.

Research in many areas can take advantage of the many common strengths of the wide range of natural history collections. Such collections can be found throughout the world, and the number of specimens in the world's largest museums exceeds 100 million each. Natural history collections have developed over the past several centuries and provide a record of patterns of biodiversity during this time frame. Many natural history collections also include fossils, allowing for the study of species through deep geological time.

Some of the most important elements of natural history collections are sets of voucher specimens.[8] These "vouchers" are specimens that are referenced to specific, usually published, research studies, and they allow other researchers to evaluate and re-evaluate the accuracy of taxonomic identifications; to identify species that were not identified when and where they were collected; or to identify cryptic species.[9] Vouchers also allow researchers to examine the patterns and causes of the range of morphological and genetic variation observed in a particular species.[10]

Natural history collections also have important limitations. Their content generally reflects interests of particular affiliated researchers, geopolitical histories, changes in scientific perceptions and funding, and other historic idiosyncrasies.[11] For example, the geographic scope of many of the collections of London's Natural History Museum mirrors the history of the British Empire; other European museums similarly illustrate their respective colonial pasts.

Natural history collections are frequently the products of individual efforts. Consequently, a particular area may be sampled thoroughly for only one or a small number of taxa. Rare species often are overrepresented in collections simply because collectors tend to be attracted to, and save, "special" finds. Most museum collections have been accumulated to address taxonomic or biogeographic questions, so specimens are retained or discarded based on individual interests, but not usually with the goal of representing the diversity of a particular habitat or region. While meeting the goals of taxonomic research, such taxon-based focus may limit the usefulness of natural history collections for research in other disciplines, such as community ecology or conservation biology.

Finally, it is important to note that natural history collections provide raw data only on the historical presence of a species. The well-known maxim that the absence of evidence should not be taken for the evidence of absence is useful to keep in mind when using natural history collections (see also Chapter 15).

NATURAL HISTORY COLLECTIONS AND THE IMPORTANCE
OF TAXONOMY

Awareness of the significance of taxonomic research and knowledge has increased dramatically over the past two decades. It is now widely recognized that only a small fraction of Earth's rich biota has been identified to date. Many fields, including agriculture, medicine, conservation, and natural resource management, depend on accurate knowledge of species identity. Incorrectly identifying species in these and other fields could result in useless activities or dangerous results.

However, at the same time that many species are threatened with extinction, the number of taxonomists who can identify known species and describe new ones is declining rapidly; this "taxonomic impediment," is recognized widely as a major issue of concern across the sciences.[12] The taxonomic impediment is central to biodiversity research in general and to studies of natural history collections in particular. We would be remiss not to emphasize that any use of natural history collections depends on sound taxonomic expertise, which is available primarily from individuals who work in and with natural history collections.

SURVEYING AND RESURVEYING THE EASTERN
MEDITERRANEAN SEA

The strengths and limitations of natural history collections when used to define and inform repeated surveys and document changes in biological diversity are best illustrated through a case study.

The establishment of non-native species is considered a major threat to marine ecosystems. Changes associated with non-native species in the structure and function of ecosystems can have socio-economic consequences that may lead to social conflicts and to economic and production losses.[13] The Suez Canal, which is an artificial, sea-level canal linking the Mediterranean and Red Seas, provides a unique colonization pathway into the Mediterranean Sea for marine organisms. The Suez Canal is thought to be responsible for turning the coastal waters of the eastern Mediterranean into the marine region that may be the most heavily affected by non-native species in the world.

Repeated surveys and resurveys of the marine fauna of the eastern Mediterranean Sea provide a reasonably accurate timeline of the colonization, establishment, and subsequent spread of marine species from the Red Sea into the Mediterranean Sea (so-called Erythraean species). This record starts at the beginning of the nineteenth century, when the French army of Napoleon I surrendered Egypt to the British. The French naturalists Geoffroy Saint-Hilaire and Alire Raffeneau-Delile, who had been studying the biodiversity of the Levant, threatened to burn their collections of the biota of Egypt if they were not allowed to keep them.[14] Saint-Hilaire and Raffeneau's collection is now preserved in the Lenormand Herbarium in Normandy, France. In fact, all the earliest natural history collections from the Levant are housed in European museums.[15]

Over the next several years, Saint-Hilaire and Raffeneau published the results of their collecting efforts and described many new species within the colossal *Description de l'Egypte*.[16] Their work is notable for (among other things) including the description of *Fucus taxiformis* Delile,[17] which is thought to be the first non-native species recorded from the Mediterranean Sea.[18] A subsequent intensive survey of the biota of the continental shelf of the state of Israel by the Sea Fisheries Research Station (SFRS) in 1946–1956 laid the basis for subsequent marine collections there. The SFRS survey resulted in a series of publications that highlighted the extent to which Erythraean species introduced via the Suez Canal had become established along the Mediterranean coast.

The specimens preserved in Israeli collections document the appearance of Erythraean species new to the Mediterranean and first noted along its coast, the establishment of reproducing populations, and the increasing abundance of formerly rare Erythraean species. The collections were enriched further with material collected by a joint effort of the SFRS, the Smithsonian Institution, and the Hebrew University of Jerusalem to investigate the spread of the Erythraean biota in the Levant and its impact on the native biota. By the end of the third year of the program, some 5,300 samples had been collected and partially sorted and identified.[19]

By the mid-1970s, it was clear that the intertidal and subtidal biota of the eastern Mediterranean Sea were undergoing rapid and profound changes, which, because of the great number of non-native species, may have had no historical parallels.[20] In the past 40 years, taxonomists and ecologists have continued to collect and record both the native biota and the spread, biology, and impact of non-native species along the Israeli coast and throughout the eastern Mediterranean. As generations of collectors before them had done, these researchers have stored their collections in natural history museums.[21] Some of the specimens are accompanied by detailed environmental data, as they were collected within the framework of pollution-monitoring surveys of the macrobenthos (see also Chapters 6–8). Overall, the detailed locality data accompanying all of these specimens provides an invaluable resource for anyone interested in resurveying these environments to understand the exceptionally rapid environmental changes occurring in the Mediterranean Sea.

IMPROVING NATURAL HISTORY COLLECTIONS TO DEFINE BASELINES AND SUPPORT RESURVEYS

Natural history collections have been especially useful as sources of data on variation in attributes of individuals, including morphology, genetic composition, and chemical make-up, associated with environmental variation. With appropriate caveats, museum specimens also can provide important information about distributions of individual species, but they rarely include additional data on local population sizes or habitat structure, which are key variables for ecological, environmental, and evolutionary research.[22]

Changes in how specimens are collected could enhance the value of natural history collections.[23] For example, specimens should be collected following particular statistical sampling designs rather than haphazardly. Additional metadata about the nature and extent of collecting effort, habitat descriptions (and photographs), and environmental conditions should be associated with specimens. When appropriate, collections should account for important environmental variables that may be used to test particular hypotheses. In addition, the increasing efforts to digitize museum collections should start by focusing on the holdings most useful for research efforts that may attract multiple constituencies. For example, in addition to digitizing type specimens and rare species, efforts should be made to digitize samples from whole-biota studies, long time series, and records of intensively sampled common taxa.[24]

Such changes will require the active participation of a wide range of researchers, not just museum-based taxonomists, in setting priorities for natural history collections. Tensions, re-allocation of funds, and refocused interests will undoubtedly result. Maintaining the delicate balance between the crucial mission of natural history collections to record, identify, and catalogue Earth's living and fossil biological diversity while simultaneously sampling with other disciplines in mind presents new challenges for museums worldwide.[25] A constructive and successful improvement of natural history collections as a basis for repeated resurveys and hypothesis testing in a range of disciplines can be achieved only if the broader scientific community recognizes the importance and long-term value of natural history collections.

NOTES

1. A few recent examples include Suarez and Tsutsui (2004); Ilves et al. (2013); Rocha et al. (2014). See also Chapters 5 and 15.

2. The literature on the relationship between biodiversity and ecosystem services is vast. For a succinct review, see Larigauderie and Mooney (2010). While some collections have benefitted from this attention (e.g., Snow 2005; Pennisi 2006; Schnalke 2011), others have not (Gropp 2003; Dalton 2007; Kemp 2015).

3. See Johnson et al. (2010) and Lister and the Climate Change Research Group (2011) for readable synopses.

4. Recent summaries are in Ponder et al. (2001); Graham et al. (2004); Pyke and Ehrlich (2010); Drew (2011).

5. For further reading about macroecology, see Brown and Maurer (1989). Macroecology provides an explicit contrast to detailed experimental studies of the myriad interactions within ecological assemblages that reveal particular mechanisms behind observed patterns but rarely yield generalities (Lawton 1999; see also Chapter 9). A parallel macro-scale approach underlies studies of macroevolution, and natural history collections are eminently suitable for such studies (see Chapter 5).

   Macroecology's demand for large amounts of data also takes advantage of the hundreds of millions of specimens in worldwide natural history collections. As informatics tools have evolved and natural history collections have been scanned and digitized (e.g., through the Global Biodiversity Information Facility: http://gbif.org), access to data in natural history collections have become available to scientists who would not be characterized as museum scientists (Johnson et al. 2010).

6. Examples of such "climate-envelope" or "niche" models are legion (see, e.g., Johnson et al. 2010; Saupe et al. 2014) and not without analytical challenges (Fitzpatrick et al. 2013).

7. The accuracy and precision of locality data vary among collections and collectors. Before the development of global positioning systems, locality most frequently was specified with a locality "name." Some of these names were

better known than others, and many were unspecified. If collectors' field notebooks are available, additional locality data may be gleaned from them (see Chapter 5 for discussion of other important uses for such notes). Likewise, contextual information that can be crucial for fine-scale resurvey/replication is frequently missing from natural history collections. In general, museum collections have been amassed by a combination of amateur and professional efforts that have emphasized taxonomy and morphology, not ecology, agriculture, or medicine, among other disciplines.

8. Although a "type" specimen is a kind of voucher, not all voucher specimens are "type" specimens. See Chapter 1 for additional discussion of type specimens.

9. For a discussion of the additional complexities of identifying the "same" place and time, see Chapter 16.

10. Common uses of natural history collections include studies of microevolution, population genetics, and chemical pollution. For example, studies of the shells of the land snail *Cepaea nemoralis* collected during the last 50 years in the Netherlands and stored in natural history collections there have revealed evolutionary changes in shell color (Ozgo and Schilthuizen 2012). The tempo and mode of changes in body size of carnivores as the climate has changed have also been explored with natural history collections (Meiri et al. 2009). As methods to extract DNA have been extended to preserved material, natural history collections have been found to be unique repositories of data on geographical and temporal variation, especially if large numbers of individuals were sampled from the same locations at multiple points in time (Wandeler et al. 2007; see also Chapter 5).

    Molecular methods also are being used with museum specimens to reconstruct the history of diseases and other pathogens (Tsangaras and Greenwood 2012). Finally, the chemical content of natural history specimens may reflect the environmental conditions when they were collected. Changes in the amount of mercury in bird specimens reflect changes in fossil-fuel emissions and other pollutants (e.g., Head et al. 2011; Evers et al. 2014). Climatic changes also can be inferred from the chemical structures of museum specimens. As temperature is one of the key determinants of the rate at which magnesium is incorporated into the skeletons of some crustose algae, increases in water temperature have led to deposition of more magnesium calcite ($MgCO_3$) in their skeletons. Reconstruction of this pattern was done with algal samples stored in herbaria (Williamson 2014).

11. Gardner et al. (2014) is a useful review.

12. The taxonomic impediment is defined by the 1992 Convention on Biological Diversity: http://www.cbd.int/gti/problem.shtml. See, for example, Jiao (2009); Patterson et al. (2010); Ebach et al. (2011); Joppa et al. (2011); Pearson et al. (2011); Bacher (2012).

13. Simberloff (2013).

14. "The Levant" was used to refer to the region of what we now refer to as the Middle East and meant lands of the Mediterranean east of Italy. At various

points in time, it included the countries of present-day Cyprus, Israel/Palestine, Jordan, Lebanon, Iraq, Syria, Egypt, and Turkey.

15. Other important collections include the MacAndrew Collection (University Museum of Zoology, Cambridge, England), which provides an invaluable source of reference relating to Mediterranean and Red Sea mollusks in the era before the construction of the Suez Canal. Collections made by the Cambridge Expedition to the Suez Canal are preserved in the Natural History Museum, London.

16. This multi-volume work was published between 1809 and 1829 and included contributions from the so-called savants: the civilian scholars and scientists who joined Napoleon I's ill-fated expedition to Egypt (1798–1801).

17. This alga was renamed 32 years later as *Asparagopsis taxiformis* (Delile) Trevisan de Saint-Léon.

18. Andreakis et al. (2007).

19. Steinitz (1970).

20. Por (1978); Galil and Lewinsohn (1981).

21. Contemporary collections from the Israeli surveys are deposited in the National Collections at the Steinhardt Museum of Natural History at Tel Aviv University and at the Hebrew University of Jerusalem.

22. Pyke and Ehrlich (2010).

23. These recommendations derive from discussions in Graham et al. (2004); Johnson et al. (2010); Pyke and Ehrlich (2010). De Carvalho et al. (2008) provides appropriate cautions.

24. See also McNutt et al. (2016).

25. Besides the obvious challenges in re-allocating or re-directing shrinking budgets, there is broad recognition that natural history collections are less "valued" than other research infrastructure (e.g. Gropp 2003; Raven 2003; Dalton 2007; Kemp 2015).

## CHAPTER FIVE

Looking to the Past to Plan for the Future

USING NATURAL HISTORY COLLECTIONS AS HISTORICAL BASELINES

Rebecca J. Rowe

Natural history collections can provide baselines of ecological conditions at decadal to centennial time scales. When paired with modern resurveys, these historical data may yield critical insights into the impacts of climate and land use on biological systems.

Anthropogenic impacts are rapidly changing climates and habitats from local to global scales.[1] As a result, forecasting how the distribution and abundance of species will change as their environment shifts is a major priority for the natural sciences. Much can be learned about how species will respond to future environmental changes by examining how they responded to past episodes of environmental change.

Specimens and associated field data stored in natural history collections represent archives of ecological information from which baseline conditions can be reconstructed at decadal to centennial time scales.[2] Pairing historical data from natural history collections with contemporary data from collections or field surveys (i.e., indirect and direct resurveys, respectively) can identify recent changes in diversity and species distribution. In this chapter, two case studies of small mammal assemblages in western North America illustrate different methods for the use of collections data in a resurvey and how the method employed is contingent upon the completeness and quality of the collections data.

COLLECTIONS-BASED RESURVEYS AS COMPLEMENTS TO MANIPULATIVE EXPERIMENTS

Fine-scale spatial and temporal studies offer a powerful approach for understanding ecological dynamics. This is particularly the case when data from observational surveys and monitoring programs are combined with manipulative experiments. Although manipulative experiments provide great insight into cause-and-effect relationships, the tractability, reliability, and minimal bias of carefully designed and replicated experiments (Chapters 9, 10) come at the expense of simplifying real-world complexity. Many ecological processes also unfold over much longer time periods than can be captured in a short-term ecological study, and short-term findings can be poor predictors of long-term responses.[3] These limitations highlight the importance of alternative approaches for understanding ecological dynamics.

Broadening the axes of space and time is especially warranted within the context of global change ecology. Retrospective approaches can identify how species and communities have responded to past changes in climate and land use and may inform on the traits and contexts that increase vulnerability.[4] In the absence of a continuous time series, however, retrospective studies require historical ecological information from which a baseline record of ecological conditions (e.g., distribution of a species or the occurrence and relative abundance of a group of species) can be reconstructed and change assessed over the interval between surveys.

The archived specimens in natural history collections provide direct evidence that a particular species occurred at a particular time and place (see also Chapter 4). The specimens and associated archival data together can document past conditions and provide a historical baseline from which to compare current conditions. Recognized long ago, this use of collections has been realized only with the passage of time and with the development of new methods of data storage and retrieval.[5]

Although still in their infancy, collection-based resurveys have been used to document species declines and changes in the distribution and composition of species assemblages for a variety of taxa at local to continental scales.[6] Natural history collections data often are used in ways that differ from the purposes under which they were originally collected. Although they are not manipulative experiments writ large, when understood and analyzed carefully, collections data and associated resurveys can provide invaluable insights about biotic responses to environmental change over time and space that are inaccessible to direct observation or experimentation.

THE NATURE OF MUSEUM DATA: SPECIMENS AND FIELD NOTES

Natural history collections were developed primarily to catalog and archive voucher specimens for systematic research.[7] Voucher specimens provide direct evidence that a particular species occurred at a given time and place. They also permit re-examination and verification of species identities, which are critical for proper reclassification pending taxonomic revisions and changes in nomenclature and for correction of errors in identification. The value of voucher specimens, however, goes far beyond confirming the existence and identification of species. Collections-based research advances our understanding of science (in particular, ecological and evolutionary processes) and has direct applications to society, including human health and sustainable agriculture.[8] Analytical and technological advances have found new uses for old specimens.[9] In concert with collections-based resurveys, these approaches provide powerful methods to develop a mechanistic understanding of the causes of changes in distribution and diversity observed over space and time.

Voucher specimens alone do not provide a complete record. Ancillary data in the form of field notes (including associated maps and photographs) compiled during the collection event can augment or enhance specimen information and can be critical for the use of collections data in ecological research. The recording and preservation of field notes, however, was not and still is not common practice. There are, of course, notable and important exceptions.[10] Under the foresight and leadership of Joseph Grinnell, the format and collection of field notes were standardized. The "Grinnellian method," complete with a journal, catalogue, and species accounts, has been used for over a century at the Museum of Vertebrate Zoology (University of California, Berkeley) and widely adopted elsewhere.[11]

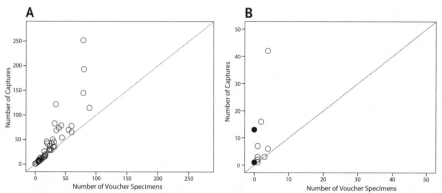

FIG. 5.1. Comparison of the total number of individuals captured and those pre-served as voucher specimens for a given species. Black circles represent species captured for which no vouchers were deposited in the collection. *Panel A:* Data from small mammal surveys in the Ruby Mountains, Nevada (1927–1929, 15 sites) and the Toiyabe Range, Nevada (1930, 12 sites). For clarity, the most common species (*Peromyscus maniculatus*) has been excluded; values are 99 vouchers from 820 captures and 50 vouchers from 664 captures, respectively. *Panel B:* Data for small mammal species surveyed at one site in the Ruby Mountains.

Field notes contain data and metadata associated with collection events. Field notes often document the numbers and identity of all individuals captured, including those not preserved as specimens. Thus, they provide a more complete record of the survey results than can be obtained from the voucher specimens alone. The discrepancy between captured and vouchered data can be great (fig. 5.1) and may have a substantial impact on estimates of distribution and diversity, depending on the scale of the analysis (see the case studies that follow). Field notes often also contain pertinent information about animal behavior, as well as the age, sex, and reproductive status of each capture (for vouchers this information may be recorded on specimen labels along with standard external measurements), from which additional demographic information may be derived.

Metadata derived from field notes may include information on location and environmental conditions, as well as survey intent and design. Details on environmental conditions may include general descriptions of the topography, vegetation, and habitat conditions; evidence of local land disturbances (e.g., non-native plant species, fire, grazing, and erosion); and summaries of local weather. These descriptions may be supplemented with drawings, maps, and photographs. Survey objectives and sampling designs are often provided and may detail information on location, devices used (e.g., trap types, firearms), collecting effort, date, and time of day.[12] The digitization of collectors' field notes is ongoing at many institutions, and these efforts, along with those that automate field-note data acquisition, should further promote their use.[13]

ACCURACY AND PRECISION OF COLLECTIONS DATA FOR RESURVEY

The quality of specimen records and field notes and the types of information that can be extracted from them will vary widely (table 5.1). The

Table 5.1

Resolution of natural history collections data used for resurveys

		FINE	COARSE
Study domain			
	Temporal snapshot	1 to a few years	Decadal to multi-decadal bins
	Spatial scale	Local	Regional
Collection events			
	Time of collection	Hours	Year
	Spatial uncertainty	Meters	Kilometers
	Survey data	All captures	Vouchers only
		Full effort	No effort
Units			
	Sites	Fixed	Not fixed
		Discrete	Pooled
	Species	Fixed	Not fixed
		Identities	Functional role or traits
	Derivatives	Activity patterns	Relative abundance
		Demographics	Rank abundance
		Abundance	Rank occurrence
		Occurrence	

case studies discussed later illustrate how collections data of varying accuracy and precision can provide a robust assessment of changes in species assemblages over time. This emphasis on the composition and structure of communities is atypical. Most frequently, collections data are used as sources of species occurrence records to model distributions and changes in those distributions over time.[14] Yet, changes in species abundances within local populations typically precede changes in the occurrence of a particular species. Thus, examination of community dynamics can yield critical insights into how biotas respond to environmental change. Natural history collections provide the opportunity to examine the response of these biotas to past environmental change and, in doing so, guide conservation and management. These efforts will benefit from a broader use of collections data.

Defining space and time for collections-based resurveys can be challenging.[15] Specimen records preserve the identity of the species and the place and time of the collection event, but the precision associated with those data is varied. Failure to recognize and account for differences in precision during resurveys can lead to inaccurate assessments of change over time.[16]

Spatial error accompanies every collecting event, modern and historical. Collaborative efforts to provide standardized geographic references for historical data, quality assurance, and estimates of precision have facilitated broader use of collections data and enabled more reliable reproduction of analytical results.[17] At the same time, it is important to recognize that georeferencing protocols follow standard rules based on the content of the textual descriptor; validation of georeferenced coordinates against the ancillary archival records (field notes or route maps) is always recommended and can identify factual errors in locality or lead to the re-association of captures with more discrete sampling sites. Mismatches in spatial precision or scale also can lead to misleading results.[18]

Similarly, the precision with which the time of the collection event is recorded can vary from perhaps the hour of day (e.g., bird counts recorded in field notes) or period (a.m. versus p.m.) to the calendar date (month, day, and year) or simply the year, the latter two of which may be the only temporal data recorded on the specimen label or in a collection's database. Not knowing the year of collection would preclude the data from being used in a resurvey context but would not diminish their use for other studies. Other aspects of collection times that affect an organism's activity and physiology, such as moon phase, seasonality, or migration or hibernation periods, should also be accounted for in resurveys whenever possible.[19]

The fundamental units of natural history collections data are the sampling sites (or localities) and the species. In a resurvey, sites or species may be either fixed or not fixed.[20] A fixed site is one where a given locality is resampled, with the same degree of spatial precision, at a later time. In contrast, sites that are not fixed are those that differ in location but are presumed to be equally representative of the previously surveyed site.[21] Both approaches are valid. The use of one over another is not dictated by the accuracy and precision of the collections data but may be influenced by those factors as well as the spatial coverage of the survey, the completeness of the survey data (e.g., vouchers versus capture data), ancillary data on site condition (e.g., land-use practices), and the study objectives. Whether sites are fixed or not fixed, they may be evaluated as individual units, or they may be pooled or aggregated over a broader domain.

The focus of comparison may also vary for the species being studied. When species are considered fixed, temporal changes in occurrence (and perhaps abundance) can be assessed for an individual species. In contrast, if species are not fixed, the unit of analysis would be an individual community or the entire assemblage, and aggregate or summary metrics would be described, such as richness, biomass, proportional diversity, or the relative representation of particular functional groups or guilds.

TWO CASE STUDIES OF COLLECTIONS-BASED RESURVEYS

Retrospective studies using natural history collections data can take many forms that can accommodate differences in accuracy and precision of data. As two case studies illustrate, there are trade-offs between the quality and completeness of the record and the spatial and temporal extent over which the data are analyzed or aggregated and the ecological inferences made. However, careful framing of hypotheses or questions and attention to detail can lead to interpretable, informative, and important results. These case studies are drawn from studies of small mammals collected from the Intermountain Region of western North America

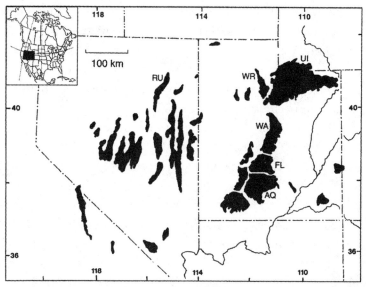

FIG. 5.2. Map of the Intermountain Region of western North America. The mountains discussed in these case studies have been labeled. The first case study from central Utah included the UI (Uinta Mountains), WR (Wasatch Range), WA (Wasatch Plateau), FL (Fishlake Plateau) and AQ (Aquarius Plateau). The second case study was restricted to the Ruby Mountains (RU) in northeast Nevada. Map modified, with permission, from that of Rickart 2001, © John Wiley & Sons Ltd.

(fig. 5.2). The high level of biodiversity of mammals in this region and its basin-and-range topography have prompted numerous surveys generating a wealth of data preserved in natural history collections.[22] Small mammals also have long been recognized as reliable indicators of environmental change because they respond to changes in climate and habitat in terms of abundance and distribution.[23] Both case studies have yielded important insights for conservation and management of mammals in this region.

Resurveys can be indirect or direct. An indirect resurvey uses compilations of paired datasets, both of which are drawn predominately from pre-existing and independent sets of natural history collections data. In contrast, a direct resurvey is one in which the modern survey(s) was done with the explicit purpose of comparing its results with parallel historical records.

An indirect approach was used to examine temporal patterns in the composition of the small mammal fauna on five mountain ranges in central Utah. Specifically, the hypothesis that recent climate warming has increased the occurrence and abundance of species adapted to dry conditions was tested.[24] The area of the study was large, and the resolution of both the raw and derived data was coarse (see table 5.1, right-hand column). The data for both time intervals were predominately drawn from museum collection databases and supplemented with contemporary field surveys conducted by me and my colleagues in each mountain range.[25]

The initial data set included 17,861 specimen records collected between 1887 and 2005. Field notes of collectors other than me were not evaluated and the spatial and temporal resolution of the specimen data varied, but the spatial uncertainty associated with each locality was known and standardized.[26] Only localities with a spatial uncertainty of less than 200 m in elevation were included.[27] As a result, only 5,294 specimen records of 27 species from 462 localities (of which my colleagues and I surveyed 65) were used in the final analysis.

The unit of analysis was the mountain range, and the five mountain ranges were considered replicates in this natural experiment.[28] Within each mountain range, localities were aggregated across time and space to establish two snapshots. The number of years in each snapshot varied, as did the interval between periods, which ranged from 27 to 53 years. By removing emphasis from any individual collecting event, these time- and space-averaged data sets provide a generalized representation of the fauna during the time period.[29]

The 27 species of small mammals were categorized by habitat affinity based on their occurrence in predominately moist habitats (11 "mesic" species) or dry habitats (15 "xeric" species). One species (the deer mouse *Peromyscus maniculatus*) was categorized as a habitat generalist.[30] The magnitude and direction of change for each species was determined and the patterns compared across mountain ranges at the level of these three habitat categories. Ranked data (for both abundance and occurrence) were used because they are less susceptible to different sampling schemes and can thus facilitate comparison when data vary in completeness and quality.[31] In examining synchrony in response among species with similar life history traits, species serve as replicates in a natural experiment (see also Chapters 9, 10). This replication improves our ability to separate the signal of environmental change from the noise of background variability, increasing the reliability and generalizability of the findings.

Marked changes in the composition of the small mammal fauna were evident and were common across the five mountain ranges. Over the past century, mesic species have become more abundant and now occur at more sites. Note that this result is opposite to that expected under climate warming; as temperatures warm and the landscape gets drier, xeric species would be hypothesized to have increased in their occurrence and abundance.[32] Examination of fine-scale observations over time at fixed sites as well as modern experiments on the impacts of grazing by livestock suggest that the increase in mesic species reflects faunal release from earlier periods of overgrazing. This interpretation highlights challenges associated with examining changes in biodiversity over the past century, a time period subject to great change in both climate and land use.

An example of a direct resurvey comes from the Great Basin Resurvey Project.[33] In this project, modern field surveys were done to compile a paired dataset on the small mammal fauna in the Ruby Mountains of northeastern Nevada (see fig. 5.1). The data were used to test hypotheses about species diversity and distributions. Both the historical and modern data are high resolution, the survey data are virtually complete, and fine spatial and temporal grain sizes (see table 5.1) were sampled within two time periods: 1927–1929 and 2006–2008.[34]

During each of the two time periods, the surveys covered the entire elevational and latitudinal extents of the Ruby Mountains, targeted the entire small mammal community, and used comparable survey methods (see also Chapter 15). Although the objectives of the survey done in the late 1920s differed from those of the one done in the mid-2000s, detailed field notes (>500 pages) made by the historical collectors allowed for accurate re-construction of survey effort and results (i.e., all capture records) for each locality. Because the old field notes included photographs and maps, it was possible to attain much greater accuracy and precision of locations of sampling sites than could be obtained from the locality descriptors on the labels of the voucher specimens and recorded in collection databases.[35]

The directed resurvey resampled 9 of the 15 historical sites, those that could be re-located with high precision and represented a range of habitat types. In addition to these 9 fixed sites, 13 additional sites (not fixed: see table 5.1) also were sampled in the mid-2000s. Both types of these replicate samples within the second time period were necessary to ensure that any pattern detected was representative of the region as a whole and was not biased by an anomalous collection event.[36] A comparison over time of abundance, biomass, and energy use for the Ruby Mountain small mammal fauna showed parallel trends in direction and magnitude when the data were pooled from all survey sites and from data that were limited to the 9 fixed resurvey sites.[37]

As in the first case study (an indirect resurvey), some findings were counter to climate-driven hypotheses. Most notably, no xeric species expanded its range upslope in the Ruby Mountains, although such a shift was hypothesized given nearly 100 years of regional warming.[38] Comparisons among functional groups showed a marked re-allocation of available resources to habitat and diet generalists and a decline across specialists (e.g., mesic and xeric species).[39] These trends among generalist and specialist species also were seen in a resurvey comparison in a neighboring mountain range; however, species-level responses can be quite variable.[40]

In summary, results from these two case studies illustrate that natural history collections can be used to facilitate comparisons at spatial and temporal scales that are inaccessible to direct observation or experimentation. These studies also underscore the conclusion that repeated resurveys using standardized approaches and done across a range of scales within and among regions can improve the inferences about how animals respond to environmental change.

ENVIRONMENTAL CHANGE AND THE VALUE OF NATURAL HISTORY COLLECTIONS

Natural history collections data may provide an imperfect match to modern sampling, but with proper attention to data quality and data coverage, and by taking advantage of new statistical methods (see Chapter 15), the vast stores of information in natural history collections can provide accurate benchmarks for assessing responses to ongoing environmental change. Two case studies from the Intermountain West of

North America have demonstrated necessary trade-offs between resolution of natural history collections data, spatial and temporal domains of resurveys, and reliability of ecological inferences. Repeatable results can be obtained from both fine and coarse scale analyses. No single spatial or temporal scale can capture processes and dynamics that occur at many scales, and increased integration of studies across scales will yield better understanding of linkages between biodiversity and environmental change, with opportunities to improve conservation and management. Natural history collections and contemporary repeated resurveys also can link historical and modern observations with results from long-term monitoring programs (Chapters 6–8) and manipulative experiments (Chapters 9–12). Modern surveys and resurveys also set and reset baselines for future resurveys. These new field studies and their associated archiving of data should be done with future researchers in mind, just as Grinnell did a century ago; continued collection and preservation of voucher specimens and the support of natural history collections are an integral part of this planning for the future. Although the specific questions that will be posed and the analytical methods that will be used in the future are unknown, preserving as complete a record today as possible will only enhance our ability to ask and answer those questions.

NOTES

I am grateful to A. Shavit and A. M. Ellison for organizing this effort and inviting my participation and to the curators and collection managers who work to preserve and facilitate the use of specimens and ancillary data. This work benefitted from thoughtful discussion with and comments by A. M. Ellison, J. L. Patton, E. A. Rickart, and R. C. Terry. The Great Basin Resurvey work was supported by National Science Foundation Grant DEB-0919409, the National Geographic Society, the American Philosophical Society, and the University of Utah Research Foundation. The Utah case study was funded by an EPA STAR Fellowship, the ARCS Foundation, the University of Chicago, an American Society of Mammalogists grant in aid of Research, and the Theodore Roosevelt Memorial Fund of the American Museum of Natural History.

1. Smith et al (2009).
2. Baseline conditions are those that were observed at the particular time at which the initial sample or survey was done. See Chapter 6 for a more detailed discussion of how the time at which a baseline was set can condition

our perception of environmental change or the potential carrying capacities or maximal harvests of particular species.

3. Kareiva and Andersen (1988) and Debinski and Holt (2000) illustrate the preponderance of short-term, small-scale field experiments in ecology and conservation biology. Good examples of the value of long-term data and insights they provide are in Hobbie et al. (2003); Jablonski (2003); Doak et al. (2008).

4. See National Research Council (2005), Dawson et al. (2011), and Chapter 4.

5. Grinnell (1910) was perhaps the first person to explicitly recognize this value for natural history collections. For discussions of the impact of the informatics revolution on museum collections, see among others Bisby (2000); Graham et al. (2004); Baird (2010).

6. The literature on this topic is too extensive to summarize here. Important reviews include Parmesan (2006), and see additional discussion in Chapters 4 and 15. Representative case studies include Moritz et al. (2008) for range shifts and Rowe et al. (2011) for community and ecosystem-level dynamics.

7. For a detailed discussion of voucher specimens, see Chapter 4. It is also important to note that documentation of ecological patterns and processes relies on species identities as basic units. Misidentifications can go undetected if voucher specimens are not collected and deposited in a natural history collection. These errors can propagate and may lead to erroneous conclusions and misguided management efforts (Bortolus 2008). A renewed appreciation for the value of natural history knowledge throughout the life sciences (Tewksbury et al. 2014), and, where permissible (i.e., when not in conflict with the study objectives or the threatened or endangered status of the study species), the collection and preservation of voucher specimens would benefit both current and future research (e.g., Moratelli 2014; Rocha et al. 2014).

8. See Suarez and Tsutsui (2004); Tewksbury et al. (2014).

9. See Nachman (2013), Wiley et al. (2013), and Chapter 4, note 10.

10. One of the most notable exceptions is the Museum of Vertebrate Zoology (MVZ) at the University of California, Berkeley. Since its inception in 1908, the MVZ procedures have considered not only systematic and taxonomic goals of collections but also broader applications in ecological research. Founding director Joseph Grinnell wrote (Grinnell 1910: 165–166): "It will be observed, then, that our efforts are not merely to accumulate as great a mass of animal remains as possible. On the contrary, we are expending even more time than would be required for the collection of specimens alone, in rendering what we do obtain as permanently valuable as we know how, to the ecologist as well as the systematist." Grinnell (1910: 166) further identified the long-term importance of the collection: "I wish to emphasize what I believe will ultimately prove to be the greatest value of our museum. This value will not, however, be realized until the lapse of many years, possibly a

century . . . this is that the student of the future will have access to the original record of faunal conditions."

Grinnell was acutely aware that voucher specimens alone did not comprise a complete record: "Our field-records will be perhaps the most valuable of all our results; hence the importance of a consistent system of record" (Grinnell 1908). This original record required data on the abundance and the extent of each species collected and an assessment of the overall species richness of the assemblage (Grinnell 1910).

11. Remsen (1977); Herman (1986).
12. As detailed in Chapter 15, such site-specific and survey-specific data can be incorporated in occupancy models to improve inferences from comparisons of historical and modern data.
13. Thomer et al. (2012).
14. See Chapters 4 and 15.
15. See also Chapter 16.
16. For additional discussion, see Chapter 4, note 7, and Wieczorek et al. (2004).
17. See Wieczorek et al. (2004) and examples at http://www.vertnet.org/.
18. For example, species distribution models may combine coarse collections-based locality data with finer-scale environmental data collected by satellites. Historical locality data may have a precision of $\geq$500 m (and upwards to tens or even hundreds of km), but the remotely sensed data often have a precision of 10–30 m. Estimates of spatial uncertainty are provided in two dimensions only; see Rowe (2005) for a discussion of uncertainty in topographically rich regions.
19. Additional implications of temporal state and migration patterns in studies of ecology and physiology are discussed in Chapter 16.
20. In parallel with definitions of terms of the general linear models used to statistically analyze these data, Shaffer et al. (1998) used the terms "fixed" and "random" for what are referred to here as "fixed" and "not fixed."
21. The decision to use sites that are fixed or not fixed may derive from more than simply spatial precision. Other factors may include the spatial coverage (completeness) of the survey data, ancillary data on conditions at that site (e.g., land use practices), and the study objectives. Even when a site is fixed and the analysis local in scale, a resurvey should not be limited to a single site, as inferences drawn at one locality over a large time gap are unlikely to yield high predictability, especially when species' responses are idiosyncratic over their range due to non-uniform changes in climate and land cover (Rowe et al. 2014).
22. McPhee (1981); Grayson (2011).
23. Hadly (1996); Thibault et al. (2010).
24. This case study is excerpted from Rowe (2007).
25. Data were primarily from the Mammal Networked Information System (MaNIS: http://manisnet.org), which has now been largely superseded by

VertNet (Distributed Databases with Backbone: http://vertnet.org). Records for Utah from the Carnegie Museum of Natural History and the Brigham Young University Monte L. Bean Life Science Museum were obtained directly from the institution.

26. All localities were georeferenced following the procedures in Wieczorek et al. (2004) as part of the MaNIS initiative and network or by me. Retrospective georeferencing is based on the verbatim textual locality descriptor, which can contain multiple sources of uncertainty that vary in magnitude and have interactive effects.

27. See Rowe (2005) for additional information on the post-hoc three-dimensional georeferencing.

28. See also Chapters 9 and 10.

29. Sites were not fixed (in the language of table 5.1), and they were pooled. By combining multiple censuses to yield time- and space-averaged data, this approach reduces the chances of misinterpreting a trend by decreasing between-sample variability and dampening the effects of potential biases in species-specific capture probabilities, incomplete survey data (i.e., vouchers underestimate occurrence and abundance), and misidentifications (see also Chapter 15). In addition, this approach ameliorates concerns often associated with the use of abundance data in temporal comparisons due to seasonal or annual fluctuations in population size (Shaffer et al. 1998).

30. Species identities were considered both fixed and not fixed (in the language of table 5.1). Because these mountains encompass a broad range of elevations that vary in conditions from dry, desert shrublands in the lowlands to moist alpine tundra at high elevations, the data were parsed to ensure there was no elevational bias in the results. Per mountain, the distribution of localities along the gradient was correlated among time periods (Rowe 2007).

31. Rank data are also more appropriate when the data were limited to voucher specimens because the "true" abundances cannot be unknown. It is notable, however, that comparisons over time using percent abundance and rank abundance showed qualitatively similar trends.

32. In fact, climate models predict this trend exactly (Cook et al. 2015).

33. Rowe et al. (2010); Rowe et al. (2011); Rowe and Terry (2014).

34. Data from the original survey was published by Borell and Ellis (1934); the resurvey was done by Rowe et al. (2010).

35. This and other directed resurvey projects, such as the Grinnell Resurvey Project (Moritz et al. 2008; Tingley et al. 2009), illustrate that historical data quality and resolution can be surprisingly good. Obtaining those data, however, requires that resurveyors use much more than the specimen records alone and evaluate collector field notes.

36. Most often, sites that are not fixed (i.e., randomly chosen) are necessary to allow for inferences across large spatial or ecological scales (see Chapters 9 and 10; for a comparable example from the health-care sector, see Finkelstein and Taubman 2015). Even when data are of high spatial resolution,

pooling or averaging across sites that are not fixed can provide a more reliable and general statement.

37. Rowe et al. (2011).
38. Rowe et al. (2010).
39. Rowe et al. (2011).
40. Rowe and Terry (2014); E. A. Rickart and R. J. Rowe (unpublished data).

Repeatable Monitoring and Observations

## CHAPTER SIX

Monitoring

REPEATED SAMPLING FOR UNDERSTANDING NATURE

Avi Perevolotsky, Naama Berg, Orit Ginzburg, and Ron Drori

Quantitative environmental monitoring programs should use replication to yield reliable and meaningful data. Adaptive monitoring programs should evolve based on an underlying conceptual framework and realistic methodologies.

The idea that it is important to conserve substantial portions of the natural environment for the benefit of people has solidified in recent years.[1] This approach calls for more protection of ecosystems and natural values as well as for good stewardship that minimizes damage to them.[2] Good stewardship of ecosystems and natural values requires knowledge of their current status and documentation of ongoing changes in them that result not only from human activities but also from natural processes. To obtain this knowledge and track changes in environmental systems, we monitor them—just as we monitor the health and well-being of people throughout their lives.

Monitoring is a regular state-of-the-system measurement.[3] When applied to natural ecosystems or species of conservation concern, monitoring is more specifically defined as the "collection and analysis of repeated observations or measurements to evaluate changes in condition and progress toward meeting a management objective."[4] A key aspect of monitoring is that the observations are repeatable and that they are done

83

at regular intervals. Such repeatability allows for the assessment of changes—relative to initial (baseline) conditions—that occur through time.[5]

Events and samples in monitoring programs must be repeated at regular intervals in defined locations. As a result, monitoring programs are beset by many of the key challenges of replication. In this chapter, we review the origin of monitoring and some of its basic principles. In the next two chapters, case studies from terrestrial (Chapter 7) and marine (Chapter 8) environments illustrate how adequate replication is incorporated into contemporary monitoring programs.

THE ORIGIN AND MEANING OF MONITORING

The word "monitoring" has its origins in the Latin word *monēre*, which means to advise, warn, or remind.[6] Indeed, one of the most important goals of ecological monitoring is to determine the state of an ecosystem at different points in time to draw inferences about changes in its state.[7] Ecological monitoring stems from the understanding that the available knowledge about nature is limited and partial and that effective conservation and stewardship may amount to no more than a continual process of trial and error. Effective management of ecological systems must be done for a long time on relatively large spatial scales. Yet, most of the available ecological knowledge is an outcome of short-term, small-scale, local efforts that can hardly tell us about large-scale, long-term phenomena.[8]

In general, the intention of ecological monitoring is to identify changes in their early stages and sound an alarm if the state of the system being monitored begins to deteriorate.[9] Ideally, monitoring provides sufficient advance warning so that effective actions can be implemented to prevent an undesired change.[10] Failing that, rehabilitation or restoration may be needed to return the component to its original state or to a desired condition.[11]

If management takes advantage of data from monitoring programs by learning, changing, and improving its actions, it is called adaptive management. If the results of each stage of management are measured and monitored, adaptations and changes to the management protocol can be developed and implemented.[12] In parallel, adaptive monitoring has been suggested as a process by which the monitoring protocol itself is modified based on new knowledge that accumulates during the monitoring program. If changes in monitoring protocols are too frequent, certain elements may not be monitored continually and it will not be possible to easily follow long-term dynamics.[13] In general, adaptive monitoring may render repeatability impossible.

There is great variance in the goals and justifications of various monitoring programs. The goals may be scientific in nature—helping to understand a phenomenon or process over space and time—or they may be management-oriented, collecting information to better inform the decision-making process.[14] But any good monitoring program must, first and foremost, address three fundamental questions: Why monitor? What is being monitored? And how is the monitoring actually done?[15]

WHY MONITOR?
There are four main reasons for undertaking monitoring programs: to inventory a system, to identify its state, to detect any trends in the state, and to determine the outcome of management actions.

Inventory monitoring is intended to broaden existing knowledge of a specific ecosystem, and it usually focuses on the presence or absence of species in the system. Inventory monitoring is also known as passive monitoring or curiosity-driven monitoring, and it has been criticized because there may be no questions or scientific assumptions guiding the monitoring program.[16] Nevertheless, inventory monitoring can be quite important because it may provide new scientific information about the ecosystem. An implicit assumption of inventory monitoring is that the accumulation of more knowledge about an ecosystem will result in more efficient conservation or management of it. However, in most cases this assumption is not tested.

State monitoring links variation in environmental or biological variables and the mechanisms or processes behind these changes (e.g., the disappearance of sand-dwelling species from coastal sands resulting from of stabilization of sand dunes and the colonization of flora and fauna from neighboring ecosystems). State monitoring is based, in many cases, on comparative measurements in which different ecological systems are monitored and compared and processes that determine the states are inferred from the observations.

Trend monitoring is used to identify directional temporal change in the system. Common indicators monitored for trends include environmental variables and population sizes of particular species. Classic examples of monitoring trends in environmental variables are the US National Oceanographic and Atmospheric Administration's monitoring of atmospheric carbon dioxide (CO_2) since the late 1950s and of the ozone hole in Antarctica since the 1980s. Methods to monitor overall biodiversity continue to evolve.[17]

Monitoring for management is used to evaluate whether a certain intervention has achieved the desired results; if it has not, the monitoring outcome may point to a better approach. Monitoring for management is a central component of overall ecosystem management. The planning and execution of programs monitoring management activities are based on hypotheses and models that predict the reaction of an ecosystem to a particular management action.[18] Surprisingly, monitoring for management remains uncommon in conservation programs. The reason may be that active management and interference with nature have not yet gained acceptance as effective and desirable conservation tools.[19] However, such monitoring ought to be developed, because conservation biologists and managers should be interested in determining the effectiveness of their management actions.[20]

WHAT SHOULD BE MONITORED?

An essential prerequisite for an efficient and successful monitoring program is determining which component of the environment or which species should be monitored.[21] In general, the things being monitored are referred to as indicators. Even the most cursory review of monitoring programs will reveal a very broad diversity of indicators, but at a broad scale, monitoring often focuses on indicators related to biological monitoring (i.e., of a particular species) or on monitoring of habitats and environmental threats.

An indicator is an entity which is indicative of another, highly important factor, and it indicates a trend or phenomenon which is not immediately and directly perceptible, and which represents monitoring goals as optimally as possible.[22] In particular, the importance of the indicator is in the additional information it communicates, above and beyond its direct measurement.[23] Thus, for example, measurement of a person's body temperature provides information about the state of the person's health well beyond the knowledge of what the actual body temperature is at the moment it is measured.

Selecting an appropriate ecological indicator is not a simple task. A good indicator needs to capture the complexities of the ecosystem yet be simple enough to be easily and routinely measured. The indicator also needs to be sensitive to the stress(ors) of interest, responsive in a predictable manner, anticipatory, and integrative; have a known response to disturbance; and have low variability in response.[24]

Most important, identifying appropriate indicators begins with the identification of a cause-and-effect model of the monitored ecosystem. It is also useful (but not necessary) to identify required and desirable attributes of indicators.[25] Among the required criteria are the coherence of the link between the indicator and what it is meant to indicate; the degree of repeatability in the use of the indicator; its measurability; and the extent of its relevance to the ecosystem. Only after an indicator has met all of the required criteria may one proceed to evaluate its fulfillment of the desirable criteria, such as rate of response to changes; sampling convenience; sampling cost; and low variance over space and time.[26]

We can also distinguish between scientific- and management-oriented indicators.[27] Scientific monitoring programs focus on the link coherence between the indicator and the state of the system. In contrast, in management-oriented programs there is more focus on the efficiency, cost, and suitability of the indicator to the monitored subject.

Biological Indicators
The use of indicators to observe variations in biodiversity is a fundamental tool for facilitating rapid action to counteract loss of species.[28] In most cases, biological monitoring tracks characteristics[29] of populations of selected species or taxonomic groups.[30] Various species-centered monitoring approaches have been proposed over the years, including focusing on common, rare, specialized, or particularly sensitive species; "umbrella" species; or functional groups. Monitoring common species may be prioritized because common species are easy to sample and may better represent characteristics of the habitat of interest than less common ones. In contrast, rare species, specialized species, endemic

species, or species that feed at high trophic levels, such as top predators, are used as indicators of system-wide change because of their sensitivity to particular environmental conditions.[31] However, such species do not always represent the overall diversity of the ecosystem, and it is usually difficult to obtain adequate sample sizes for robust statistical analyses. Furthermore, as more attention is paid to those ecosystem components that provide services for people, monitoring has emphasized more common or widely used species.[32]

Alternatively, so-called umbrella species are monitored in many conservation programs.[33] Because umbrella species require large, contiguous (unfragmented), and high-quality habitat, the idea is that if an umbrella species is doing well, other species common to the same habitat also are likely doing well, and the conservation of the habitat of the umbrella species will necessarily provide protection for the other co-occurring species.

All of these examples depend on monitoring a particular species. However, the use of specific indicator species (aka key species or focal species) has been criticized on both theoretical and practical grounds.[34] As a result, monitoring of functional groups (groups of often unrelated species with similar ecological functional characteristics) or groups of specialized species has become a central target of monitoring.[35] At the other extreme, monitoring of overall biodiversity presents greater challenges, many of which are discussed in more detail in the case studies (Chapters 7, 8).[36]

Environmental Indicators of Habitat Quality

Climatic, chemical, and physical conditions are often used as environmental indicators of habitat quality. Monitoring characteristics of habitats, rather than species, is based on the assumption that organisms are connected to the habitat they inhabit, and therefore they will react to changes in it. Monitoring habitats often involves advanced mapping methods or computer programs that yield quantification of landscape structures and characteristics that species can use (i.e., their ecological niches).[37] This approach works well for species that are habitat specialists but is less effective for generalist species that inhabit a variety of habitats. Finally, although environmental indicators usually are less expensive to monitor than biological indicators, they often do not explain trends in biodiversity very well.[38] Environmental indicators may be suitable for initial classification of habitats but cannot be a substitute for more precise measurement of diversity, which should be obtained through biological indicators.

HOW TO MONITOR

There are many different ways to construct monitoring programs, because ecosystems are complex. There is a broad consensus that monitoring should be issue-, problem-, or goal-oriented, and developing programs to meet these goals requires a conceptual model that identifies the key issues, problems, and goals.[39] However, the role of conceptual modeling as an essential stage in designating a monitoring scheme is sometimes ignored. We suggest here a set of conceptual models that can aid in the development of monitoring programs.

The assumption that most observed changes in a particular indicator originate from a single or multiple stress factor suggests a stressor-based model. Such a model places the focus of monitoring on the (often external) source of the stress (e.g., a pollutant) rather than on a species- or system-specific variable that is affected by it. In many cases it is easier to monitor an external stressor than it is to monitor the changes it causes on the system. Further, monitoring the stressor may expose processes that are taking place behind the scenes (e.g., the response of trees to acid rain or long-term drought).

In other cases, the interest is in monitoring the ecological state of a specific area or landscape, and, accordingly, a program to monitor the state of nature would be developed. For example, the US National Park Service developed a monitoring model to determine the status of and trends in the conditions of each park's priority resources and values.[40] This ambitious program deals with nearly 300 parks located in a wide range of environments. The plan includes a clear definition of the programmatic goals; collection of existing relevant data; development of a conceptual model of the ecological status in each park; selection and prioritization of indicators; formation of monitoring protocols; building a sampling system; setting up a system for processing the data, and forming a documentation path. See Chapter 7 for a comparable case study from Israel's Biodiversity National Monitoring program.

Actually monitoring something includes several components, each of which must be repeatable: a sampling protocol, a monitoring methodology, and a workflow for handling and analysis of the data.[41] For sampling, significant emphasis is placed on the duration of the monitoring program, which is related to its ability to deliver results that are of management or scientific value. The monitoring methods themselves should be consistent and provide reliable quantitative data. Strong monitoring methods that provide such reliable data take adequate time, labor, and money and often must be repeated regularly for many years. Although monitoring methods that provide qualitative rather than quantitative data are perceived as being less accurate or robust, they can be less expensive and easier to use; qualitative data gradually are finding a place in many monitoring programs. A recent compromise is the use of cameras or remote sensing devices, which provide both qualitative and quantitative data at relatively low cost.[42]

Finally, it is important to keep in mind natural variability and the limitations of any monitoring program. Most ecosystems are spatially quite variable, and the design of any monitoring program and analysis of the resulting data must account for differences in microclimate, natural habitats, and terrain units, as well as land-use history.[43] Further, not everything chosen for monitoring is equally detectable, and many taxonomic limits of many species are poorly understood.[44] These and other challenges are explored in additional detail in the next two case studies.

NOTES

We thank Yonathan Shaked for helpful comments on an early draft of this chapter.

1. This idea, currently called the "new conservation," has been articulated most clearly by Kareiva and Marvier (2012); Marvier (2012). However, there remains a serious and unfortunately acrimonious debate concerning the relative merits and sustainability of the "old conservation" (conservation based on the intrinsic value of nature) and the "new conservation" (Minteer and Collins 2008; Minteer 2012; Doak et al. 2014; Tallis and Lubchenco et al. 2014; Minteer and Pyne 2015) and whether the "new" conservation is even really that new (Farnsworth 2014).

2. Field et al. (2007).

3. McGeoch et al. (2011: 3).

4. Elzinga et al. (2001: 2). See also Lindenmayer and Likens (2010b) for a more detailed discussion of ecological monitoring.

5. Baseline conditions are those that were observed at the onset of monitoring. If monitoring had started at a different time, the baseline conditions likely would have been different. Our perception of the state of the environment may also be conditioned by the baseline. For example, most fish stocks are currently considered to be overfished, but defining the sustainable population size of any fishery depends, among other things, on knowing the carrying capacity (or maximum possible population size) of the species. Since monitoring programs rarely began before the mid-twentieth century, our estimates of carrying capacities from, say, 1950 reflect several hundred years of prior fishing for which we have less reliable data. As historical data accumulate, we may alter our idea of the appropriate baseline. This "shifting baseline" problem is a central conundrum for managing ecosystems and avoiding their collapse (e.g., Jackson et al. 2001; Lotze and Worm 2009).

6. *Oxford English Dictionary* online (http://www.oed.com/), accessed 12 January 2015. For further discussion, see Elzinga et al. (2001).

7. In most cases, only one or a few of the key environmental conditions or individual species, not the entire ecosystem, are actually monitored. This is because we rarely have the capacity to monitor all aspects of an entire system, so we normally monitor particular *indicators* that can provide information about unmeasured (or unmeasurable) conditions. See Suter (2001); Lindenmayer and Likens (2011). For brevity, however, we refer to the "ecosystem" as the object of monitoring. See also Yoccoz et al. (2001).

8. See Kareiva (1989) and Stevens (1994) for a discussion of the tension between small-scale data and large-scale management needs.

9. See Dale and Beyeler (2001); Yoccoz et al. (2001); Beever (2006). Of course, monitoring could also reveal changes for the better.

10. Contamin and Ellison (2009) illustrated that even for ecological systems for which we have deep understanding of their structure and function, many decades of advance warning may be required for preventative actions to be effective.

11. Systems may also recover on their own, even in the absence of human intervention. See Foster and Orwig (2006) and Jones and Schmitz (2009) for salient examples.

12. Holling (1978) introduced the concept of adaptive management to ecologists and environmental scientists. Elzinga et al. (2001), Yoccoz et al. (2001), and Beever (2006) discuss the importance of monitoring data for adaptive management. See Ringold et al. (1996) and Stankey et al. (2006) for two useful examples.

13. Beever (2006); Lindenmayer and Likens (2009).

14. Yoccoz et al. (2001).

15. See Legg and Nagy (2006); Lindenmayer and Likens (2010a).

16. Lindenmayer and Likens (2009).

17. Jones et al. (2011).

18. Nichols and Williams (2006).

19. Stankey et al. (2006).

20. Sergeant et al. (2012).

21. Jones et al. (2011).

22. Hammond et al. (1995); Heink and Kowarik (2010).

23. Niemeijer and de Groot (2008).

24. Dale and Beyeler (2001). Müller (2005: 281) encapsulates these criteria in recommending that a good indicator provides "a general quantifying representation for the integrity of ecological systems."

25. For example, when monitoring top predators as indicators of overall biodiversity, it may be desirable to begin by creating a model of the system's food web and the trophic relations between the organisms within the area to be monitored (O'Brien et al. 2010). See Belnap (1998) for required and desirable (but not necessary) attributes of indicators.

26. There is extensive literature on the ideal requirements of a successful biodiversity indicator. For example, McGeoch (1998) provided a list of 23 criteria for biodiversity indicators, and Schmeller (2008) proposed a partially similar list of criteria for broader classes of indicators.

27. Heink and Kowarik(2010).

28. Lindenmayer and Likens (2011).

29. Such characteristics include species composition and demographic variables such as population size, sex ratio, and juvenile-to-adult ratio, all of which give an idea of the status of one or more species under certain environmental conditions in a particular place at a particular time.

30. In many biodiversity-monitoring programs, two opposing approaches have dominated. One advocates monitoring as many species as possible, whereas the other advocates a focus on particular indicator species (Lindenmayer and Likens 2011). The attempt to monitor as large a number of species as possible often comes at the expense of other aspects of a monitoring program (e.g., Heink and Kowarik 2010). On the other hand, we understand today that there is no one species or one taxonomic group that is the object of the "perfect" monitoring scheme. Often the most appropriate group is the one that is most closely tied to management goals.

31. Beever (2006).

32. Kremen (2005); Obrist and Duelli (2010).

33. Beever (2006).

34. The general argument against single-species monitoring is that a single species will not withstand the test of a sound monitoring program and cannot reflect overall ecosystem complexity. Rather, monitoring several species together will yield better insight. General critiques are provided by Simberloff (1998); Lindenmayer et al. (2002); Lindenmayer and Likens (2011); Sætersdal and Gjerde (2011).

35. The flora is the cheapest biological indicator that provides a good representation of over 70 percent of the variance in diversity (Mandelik et al. 2010). Kati et al. (2004) present similar findings regarding monitoring woody vegetation alone.

36. Even the term "biodiversity" has many meanings. It is used to discuss diversity at all levels of biological organization: genetic diversity within populations; species diversity; diversity and heterogeneity of habitats; life forms; and ecological processes (Schmeller 2008). Monitoring biodiversity, therefore, first requires a decision about what type of biodiversity to monitor (Teder et al. 2007). Second, monitoring biodiversity needs to be constrained by area, habitat, or even particular taxonomic groups (Parkes et al. 2003). Drori et al. (Chapter 7) discuss how choices were made in Israel's Biodiversity National Monitoring program to focus on flora and fauna that characterize dynamic ecological states.

37. McGarigal and Marks (1995) developed one of the first software systems for analyzing habitat fragmentation, and it continues to be used for analyzing data from monitoring habitats. More recently, a large-scale attempt was made to formulate an outline for uniform monitoring of habitats based on landscape ecology. The outline focused on the structure and spatial relationships of the physical and biological characteristics (Bunce et al. 2008).

38. Mandelik et al. (2010).

39. Fancy et al. (2009) stress that these models integrate existing knowledge on the ecosystem to be monitored; indicate important processes; and imply connections between potential indicators and ecological states or processes. See also Lindenmayer and Likens (2010a, 2010b).

40. Fancy et al. (2009).

41. Vos et al. (2000); see also Chapters 13 and 14.

42. Ahumada et al. (2011).

43. See, for example, Buckland et al. (2005).

44. Sarmento et al. (2011).

## CHAPTER SEVEN

Monitoring the State of Nature in Israel

Ron Drori, Naama Berg, and Avi Perevolotsky

Effective monitoring of biodiversity uses replicable methods to sample key indicators of ecological threats and processes in discrete ecosystems and regions. At repeated intervals, monitoring produces an assessment of the "state of nature." Conceptual models of environmental drivers and responses are developed for each ecosystem or region to link the indicators and assessments to conservation or management goals.

We present a case study of a national monitoring program encompassing a diverse set of ecosystems. Israel is located at a crossroads of Africa, Asia, and Europe. It sits at the southeastern tip of the Mediterranean ecosystem, bordering the vast Saharo-Arabian desert belt to its south and connected via the Rift Valley to the heights of Southeast Asia and the dry tropical ecosystems of East Africa. This combination of geography and ecology provides habitats for a remarkably high number of species, but the high densities of Israel's population and its rapid development, intensive land use, and climatic change threaten this biodiversity.[1] As discussed in Chapter 6, a key component of a strategy to conserve biodiversity is a monitoring program that can identify the current state and trends—stable, declining, or thriving—of biodiversity in a country. In this case study, we discuss the goals and implementation of the Israeli National Biodiversity Monitoring Program (IBM), with particular attention to the challenges of replication and repeatability in this long-term monitoring program.

GOALS AND DESIGN OF THE IBM PROGRAM

The hypothesis that the IBM is meant to test is that the state of nature in Israel is gradually deteriorating as a result of rapid development and urbanization processes, the diminishing of habitats, and ongoing climatic changes. To test this hypothesis, the IBM aims "to assess the state of nature in Israel in order to identify significant changes, especially those that express degradation and damage to the biodiversity and its functions (at various levels—species, ecosystems, landscape), in an effort to propose ways of halting such processes."[2] Three basic principles were incorporated into the IBM: (1) emphasize monitoring the state of nature, not the outcome of management activities (the latter should be done by lead organizations on a site-by-site basis); (2) use robust scientific tools to the extent possible; and (3) commence with an appropriate conceptual model (see Chapter 6 for a discussion of such conceptual models) specifically tailored to each relatively homogeneous region or ecosystem (hereafter "unit") that is to be monitored.

Ten regional, relatively homogeneous monitoring units were identified that cover most of the state's countryside (fig. 7.1). Following the first principle, urban and infrastructure areas were excluded. Agricultural lands and aquatic habitats also were excluded from the program, although specific monitoring schemes have been designated for them (see a detailed example in Chapter 8).

Each monitoring unit was defined and characterized by a dedicated team of experts.[3] Each of these "unit-teams" was responsible for the following:

1. Defining the monitoring unit and characterizing its main ecological features and its spatial distribution throughout Israel.
2. Identifying a list of main threats and processes (THPs) occurring in the monitoring unit. Each THP was inferred to be a driver that strongly affects the biodiversity or functioning of the unit and that it is important to monitor over the long term. The team members also ranked the THPs according to their importance in determining the state of nature in the unit.

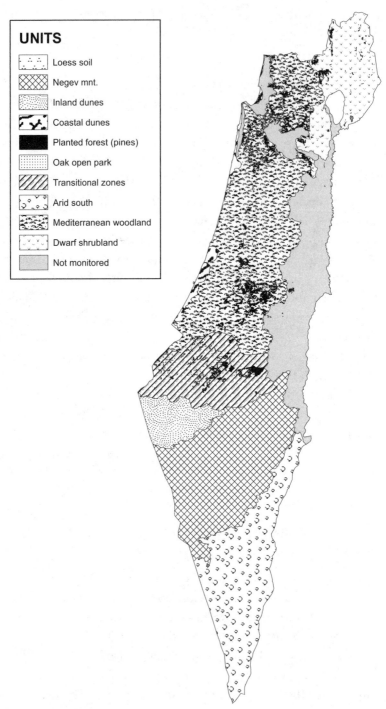

UNITS

- Loess soil
- Negev mnt.
- Inland dunes
- Coastal dunes
- Planted forest (pines)
- Oak open park
- Transitional zones
- Arid south
- Mediterranean woodland
- Dwarf shrubland
- Not monitored

FIG. 7.1. The division of Israel into monitoring units. Urban areas and agricultural land were excluded from the map for reasons of clarity.

3. Developing a list of indicators to be monitored for each THP. As described in Chapter 6, the indicators identify, in light of one or more THPs, the state of the unit or key species and processes occurring within it.

4. Recommending a preferred methodology for repeatedly measuring and monitoring the indicators.

The definitions, THPs, and monitoring recommendations for each unit were then reviewed by an evaluation panel.[4] The evaluation panel generally did not interfere with the scientific considerations and conclusions of each unit team but did address some logistical considerations, such as organization and costs, which the original teams had been instructed to ignore. We provide an overview of the different monitored units, the corresponding THPs for each unit, and the related indicators for each THP (table 7.1). The IBM began in early 2011, and the first monitoring cycle was completed by the end of 2014.[5]

BUILDING IN REPLICATION

The importance of spatial and temporal repeatability was considered throughout the design (mapping and planning) of the IBM. Existing spatial knowledge of each unit, especially of THPs that were intended for monitoring, varied from one unit to another. In certain units, such as the planted forest, there were reliable and spatially precise maps of forest stands. In other units, such as the Mediterranean woodland (maquis), there was general spatial knowledge of their extent but no precise maps of their spatial extent or location.[6] Other units, including semi-arid dwarf shrublands and the semi-arid eco-transitional region, are not spatially contiguous units but rather occur on the landscape as fragmented patches. Aggregation of these patches into monitoring units required development of new mapping solutions.

The distribution of sampling sites within and among units was determined by the nature of the THPs of interest. For example, factors with impacts at local scales, such as rural villages and agricultural fields, were thought to have strong impacts on their immediate surroundings; the intensity of these impacts was modeled as a diminishing function of distance from them. Consequently, monitoring local-scale factors was accomplished by adding spatial replicates located at fixed distances from

Table 7.1

An overview of the Israeli Biodiversity Monitoring program: Spatial units, threats/processes, and indicators

ECOSYSTEM / GEOGRAPHIC UNIT	THREATS/PROCESSES TO BE MONITORED	INDICATORS TO BE MONITORED
Mediterranean woodland (maquis)	Effects of agricultural villages	Changes in mammalian, butterfly, and avian communities
	Woodland closing up and pine colonization	Changes in woody cover and composition; presence of pine seedlings; changes in mammalian, butterfly, and avian communities
	Demography—Recruitment of oak seedlings	Oak seedling density
	Climate changes	Woodland desiccation (percent of dry individuals)
Oak (*Quercus ithaburensis*) open forest	Dynamics of woody vegetation	Changes in cover and composition of the woody vegetation; changes in mammalian and avian communities
	Livestock grazing	Changes in the cover and composition of the woody vegetation; oak seedling density; changes in mammalian communities
	Effects of agricultural villages	Changes in mammalian communities
	Oak recruitment and mortality	Oak seedling density; ratio of dead/dry individuals in the sampled population
	Effects of recreational activities	Changes in mammalian and avian communities
Planted conifer forests (mostly pines)	Forest crowding	Changes in mammalian, butterfly, reptile, and avian communities; cover and composition of the woody understory vegetation

	Climate change	Physiological status of the stand; ratio of dry trees; density of newly established oak seedlings
	Effects of recreational activities	Changes in mammalian and avian communities
Dwarf shrubland	Effects of agriculture	Changes in mammalian and avian communities
	Effects of habitat fragmentation	Changes in mammalian and avian communities; changes in woody vegetation formation
	Livestock grazing	Changes in woody vegetation cover
	Successional processes	Changes in woody cover; changes in mammalian communities
Coastal dunes	Urbanization impact	Woody and herbaceous plant communities composition; changes in reptile and rodent communities
	Invasion by exotic plant species	Changes in rodent communities; invasive plant species cover; presence of highly invasive, exotic herbaceous species; changes in reptile communities
	Dune stabilization	Changes in rodent communities; woody community composition; herbaceous: woody ratio with emphasis on invasive species; changes in reptile and beetle communities
Semi-arid eco-transitional region	Climate change	Changes in reptile and avian communities; changes in woody cover and composition, including mortality; primary production
	Urbanization impact	Changes in mammal, avian, and reptile communities
	Grazing	Changes in woody cover; changes in mammal, avian, and reptile communities.

(continued)

Table 7.1 (Continued)

ECOSYSTEM/ GEOGRAPHIC UNIT	THREATS/PROCESSES TO BE MONITORED	INDICATORS TO BE MONITORED
Loess plains (Northern Negev)	Runoff catchments	Changes in avian and reptile communities, including nest densities; changes in cover of woody and herbaceous vegetation
Negev Heights	Effects of settlements	Changes in mammalian and avian communities; isopod burrow densities; woody vegetation cover and composition
	Climate change	Changes in woody vegetation cover and composition, including mortality
	Grazing	Changes in woody vegetation cover and soil properties
Inland dunes	Soil crust expansion	Changes in rodent, reptile, invertebrate, and avian communities; changes in cover and composition of the woody vegetation; changes in herbaceous biomass and diversity
	Effects of agriculture	Changes in avian, reptile, and beetle communities; changes in woody cover and composition; changes in herbaceous plant diversity
	Effects of settlement and infrastructures	Changes in mammal, avian, and reptile communities.
	Climate change	Woody vegetation cover and mortality; primary production
The arid south	Effects of agriculture	Changes in mammalian and reptile communities; changes in woody cover and composition; monitoring of acacia trees density and mortality
	Climate change	Changes in woody cover and mortality; monitoring of acacia trees density and mortality

the source of the disturbance. In contrast, regional-scale factors, such as climatic changes, were thought to influence processes well beyond the unit actually being monitored and were assumed to be spatially uniform at each point within the sampled space.

Within each unit, additional spatial replicates—sites—were established for sampling and repeatedly monitoring each indicator. Plots sampled within sites were of variable shapes or of fixed areas but without predefined shapes and were located at specified points in the unit.

CASE STUDY: MONITORING THE MEDITERRANEAN MAQUIS ECOSYSTEM

The Mediterranean maquis unit was defined as the dominant natural woody vegetation formation growing in the Mediterranean region of Israel, comprised of areas receiving >400 mm of precipitation each year and with at least 100 Mediterranean trees per hectare. The maquis unit was further subdivided into three zones based on annual precipitation: areas receiving 600–800 mm, 450–600 mm, and 350–450 mm/year, respectively.

The four THPs selected for monitoring in the maquis unit were as follows: (1) successional dynamics of woody vegetation, including landscape closure by woody vegetation (an outcome of the succession process) and colonization of pine trees; (2) impact of nearby rural villages and agriculture; (3) recruitment and mortality (i.e., demography) of oaks (the most common tree genus in the maquis); and (4) climatic change represented by desiccation and subsequent mortality of mature woody plants.

Replicate rural villages with similar anthropogenic characteristics that were surrounded by maquis were chosen so as to allow for sampling of similar ecological impacts. In this case, a team of experts recommended choosing rural villages with poultry enclosures. The size of the uninterrupted maquis around relevant rural villages in each region was examined after the removal of all possible disturbance sources (trails, structures, and agricultural areas) using tools in geographic information systems (GIS), aerial photographs, and ground survey (fig. 7.2, *panel B*). At the end of this process a limited number of rural villages were left for monitoring, and five villages were randomly chosen in each ecological zone.

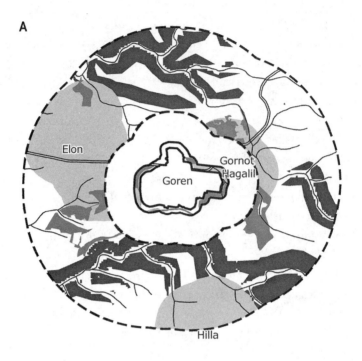

A

Elon

Gornot Hagalil

Goren

Hilla

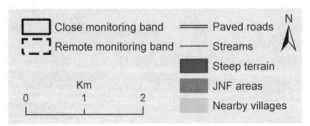

Close monitoring band   ===== Paved roads

Remote monitoring band   —— Streams

Steep terrain

JNF areas

Nearby villages

N

Km
0   1   2

FIG. 7.2. Monitoring areas (A) and mapped monitoring plots and sampling points (B). *Panel A:* Solid lines and dashed lines, respectively, delineate the bands defined for sampling close to and remote from a rural village. The various polygons represent disturbed lands that cannot be sampled and thus were excluded from the design. JNF = forests planted by the Jewish National Fund. *Panel B:* Areas to be monitored are in white, and shaded polygons represent potential habitats. Point counts were done at the center of each polygon, and vegetation sampling plots were randomly located in each polygon. The black circles mark the locations of camera traps.

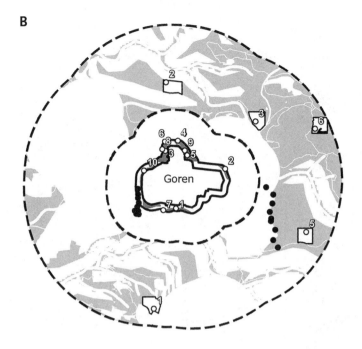

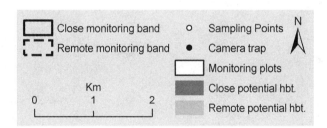

Each site was divided into three bands defined by their distance from the perimeter of the disturbance source (e.g., rural village). The first band contains the area between the perimeter of the rural village and an edge 100 meters from the perimeter; the second band extends between 100 to 500 m beyond the perimeter of the rural village; and the third band, defined as being remote from and less affected by the local-scale factors, reached from 500 to at most 2000 meters from the village perimeter (see fig. 7.2, *panel A*). For analysis purposes, spatial replicates within the maquis are the bands of each rural village that are located at similar distances from their perimeters.

Table 7.2

Principal indicators for monitoring corresponding to the threat/processes defined for the Mediterranean maquis (woodland) and the proposed methods of monitoring them

THREATS/ PROCESSES	INDICATORS	MONITORING METHODOLOGY
Successional dynamics of woody vegetation	Cover of woody plants and cover of open space	Remote sensing—mapping of woody vegetation and estimation of its cover using image processing methods
		Field monitoring of plots along fixed transects; cover estimation of each woody species
	Composition of woody species	Extraction from transect data
	Presence and cover of pine seedlings	Remote sensing—separation between broadleaves and conifers and estimation of the percentage of coverage by pines using infrared aerial photography
	Composition of avian community	Point counts—recording all of the species and the number of individuals observed and heard within plots of fixed radius
	Composition of mammal community	Fixed transects with camera traps for monitoring large mammals; rodent traps for small mammals
	Composition of butterfly community	Fixed transects in the natural maquis along which all butterfly species, as well as the number of individuals, were recorded

Rural village impact	Composition of the avian community	Point counts
	Composition of the mammalian community	Fixed transects with camera traps; rodent traps
	Composition of the butterfly community	Fixed transects
	Recruitment of oaks and pines—seedling density	Extraction from transect data (only far away from village)
Demography	Mortality and desiccation of seedlings and of mature oaks and pines	Extraction from transect data (only far away from village)
Climate change	Cover of woody plants and cover of open spaces	Remote sensing using satellite images.
	Physiological status of the woody vegetation	Estimation of the NDVI (normalized difference vegetation index), an index of the physiological status of the woody vegetation

We have summarized the indicators monitored at each sampling site within the maquis unit (table 7.2). At each site, and within the first (<100m) and third (>500m) sampling bands, three sampling plots (pseudo-replicates; see Chapter 9 for details) were established for sampling each indicator (see fig. 7.2, *panel B*).

Sample methods differed for each indicator (see table 7.2). For example, when monitoring the impact of rural villages on the number of bird species and their abundance in the maquis, counts were made at three randomly located points in the first band and at three randomly located points in the third band. At each point, these "point counts" were made in a 6.25-ha polygonal plot centered on the point.

Vegetation composition was sampled in plots that overlapped with locations where birds were counted to allow for joint analysis of the data. The habitat of each sample plot was characterized with aerial photography. Vegetation was sampled at random locations within the plot. Although the vegetation-plot area was fixed, the number of plots increased with the density of the vegetation, whereas their size decreased with vegetation density.[7]

Sampling vegetation in small plots allows for precise measurement of stem counts and for fine-scale differentiation of individuals and species. However, time and personnel were not unlimited, and the number of small plots sampled could not cover a large area (i.e., more than hundreds of square meters). Further, after several experiments in the field it became apparent that different surveyors obtained different results for the same plots, limiting the reliability of these small-scale samples.

As an alternative to small-scale hand samples, remotely sensed digital color aerial photographs were used to map vegetation cover for the entire country.[8] Custom software was developed to identify vegetation types based on reflectivity at half-meter and 10-meter resolution (fig. 7.3). This method can be repeated to evaluate changes in vegetation cover over time.

Finally, the researchers intended to sample mammals in the same plot where birds and vegetation were sampled. However, the unit teams and evaluation committees could not agree on a how to share the location of the monitoring plot, which led to the separation of mammal monitoring from the monitoring of other indicators.[9] To monitor mammals with respect to the rural village THP, camera traps were set up in sample

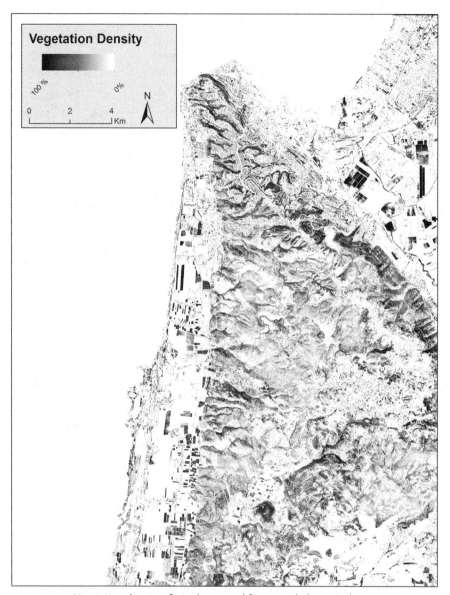

FIG. 7.3. Vegetation density of Israel extracted from aerial photography.

plots within the inner and outer sample bands.[10] In each band, nine cameras separated by 100 steps (approximately 100 meters) were deployed for 10 days each.[11] After 10 days, any captured images were downloaded and the cameras were moved to another site. Although sites in each unit were not sampled concurrently because of limited numbers of cameras, they were sampled during the same season, which was considered a temporal replicate by the researchers.[12]

DISCUSSION

Scientific research in general and ecological research in particular attempts to understand processes through replicated experiments in which it is possible to control hypothesized causal variables (see Chapters 9, 10). Monitoring, on the other hand, uses repeatable observations to document processes and trends by sampling state variables that cannot be controlled by the researcher (see Chapter 6). The salient difference between observational monitoring and controlled experiments is that much greater attention must be given in the former to selection of indicators and to precise specification of methods and sites for samples or surveys repeated in time and space.

There are interrelationships and trade-offs between selection of indicators and specification of repeatable methods. Although it is not possible in a monitoring program to control experimentally the different variables related to the monitored process of interest, there are at least two possible ways of capturing the true "signal" of cause and effect from within the "noise" of background variability: (1) selection of similar sites to sample and (2) selection of one or more indicators whose behavior depends most strongly on the underlying (and unsampled) process of interest.

It also is important to keep in mind that several processes (THPs in the IBM) are investigated and monitored in each unit. For efficiency and to control costs, and also to examine covariance statistically, the different indicators often are sampled at the same sites. Although it is not possible to select sampling sites (for example, rural villages affecting ecologically their surroundings) with identical characteristics, in most cases, sites with quite similar characteristics (e.g., the socioeconomic status and/or ethnicity of rural villages) can be identified. However, similarity between sites is likely to be different for each indicator, and "similar" sampling sites for one indicator may not necessarily be considered similar for a different indicator.[13]

Therefore, determination of sites for monitoring a phenomenon reflects the selection of the indicator to be sampled, and vice versa. Further, fixing sites for one process influences the selection of the indicator for other processes. The question therefore arises as to where to start: should the sites be selected first and only then the indicators, or vice versa?

In the IBM program, indicators were selected first based on the principal processes and threats that were considered to be the main drivers of ecosystem processes of interest. Subsequently, however, an iterative process was used to evaluate the question of the suitability of the sites for the indicator and the indicator's suitability for the site. In selecting the locations of sampling points, there were greater degrees of freedom, and a sampling point could be tailored to the indicator at more than one location. For example, sampling points for birds were chosen from areas that were defined a priori by ornithologists as suitable. Similarly, camera traps for sampling mammals were set up according to the discretion of the zoologists studying the mammals (see also Chapter 5).

Another point to consider was the influence of THP definition on available sites. For example, zoologists urged the selection of contiguous, un-fragmented forest stands > 100 hectares in area in which to monitor mammals in planted forests. Appropriate stand size was derived from characteristics of the monitored mammals (related to habitat size and home range) and the quantity and distribution of the cameras traps. However, analysis of forested areas did not produce sufficient contiguous pine stands of the specified ages and sizes. Even when the required stand size was reduced by half, to 50 hectares (ha), the researchers were still unable to find more than five units for monitoring. They therefore compromised on this reduced sample size, understanding that they would not be able to characterize the use of a forest in smaller units. Even so, there were only a very small number of sites that met these criteria, even before other sampling constraints were considered.

From this experience establishing a monitoring program in Israel, we conclude that in large-scale monitoring programs, unlike in controlled experiments, the ability to identify replicate sites and establish a repeatable sampling program was limited and depended on many seemingly independent decisions made during the program design. There also were continuous interactions and compromises among site selection, processes of interest, and indicators to be monitored. The subsequent influences of the choices and decisions made during the design process rarely were clear but imposed difficulties during program execution. Because the IBM aims to monitor processes rather than specific indicator variables, the focus was on replicating the process or threat. Replicating the process impact imposed limitations on indicators and site selection. Iterative steps refining the indicators and identifying sites helped to achieve monitoring goals in the final programmatic design.

NOTES

1. Dolev and Perevolotsky (2004); Shmida and Pollak (2007).
2. Berg and Perevolotsky (2011: 9).
3. In total, over 80 experts participated in the project, including scientists, managers, and field personnel in various disciplines who were employed by governmental agencies or academic institutions.
4. The evaluation panel consisted of the chairs of each unit team, the project directive committee (consisting of representatives of the organizations in charge of managing open landscapes in Israel), and other scientists and experts. The panel studied the team's recommendations in depth and assessed the program through four different stages: (1) Discuss, modify as needed, and approve the characterization of the proposed monitoring units. (2) Select the four most important THPs from a longer list proposed by the initial unit team based upon the original order but also on how the THPs matched or complemented those of other monitoring units. These four THPs from each unit comprised the final monitoring program. (3) Select a final list of indicators for each THP included in the monitoring program. (4) Establish a basis for unifying and integrating the various units into a nation-wide integrated monitoring program through a few common THPs. Accordingly, the panel decided that the impact of rural villages/human activities and climate change will be monitored in all 10 monitored units so as to yield a country-wide perspective that will parallel data collected from individual units. As a result, certain indicators were added to some units to ensure complete spatial replication for that indicator.
5. Budgetary limitations and the need to build up detailed protocols determined the pace of the first round of monitoring. Complete analysis of the data was completed during 2015. Based on lessons learned in the first round, the second round of monitoring, including complete data analysis, is expected to take not more than two years.
6. That is, geographic information system (GIS) overlays for these types of units were unavailable.
7. Specifically, plots to be monitored in an "undisturbed" area were identified using the Generalized Random Tessellation Stratified procedure (Stevens 2004) for spatially balanced sampling. First, ten potential plot centers were identified (only three were actually used; the additional points were located for potential future sampling). Then, around the center of each plot center, a 250×250-m square was placed. Any areas within this square classified as

"disturbed" were deleted from the sampling area. If the remaining area was sufficient for monitoring, the process was stopped and the polygon became a candidate for sampling. Otherwise, a new, larger square was placed and the filtering for disturbance was repeated. This overall process was itself repeated until 10 polygons were defined for potential monitoring. Not every polygon necessarily had the required area, and undisturbed areas could be fragmented. Finally, sites for which not enough monitoring plots of the required size could be identified were removed from the potential list of monitoring sites within the monitoring unit.

8. Remote sensing—the use of satellite, aircraft, or terrestrial sensor data— yields large numbers of repeated samples over large areas (e.g., Duro et al. 2007). Although the spatial resolution of on-the-ground sampling by hand usually is higher than that achieved by remote sensing, sensors may capture more of the variability of the system (Cabello et al. 2012). Furthermore, monitoring through remote sensing relies for the most part on only one instrument or a small number of instruments, each of which usually has a known and uniform level of measurement error. In contrast, monitoring by people in the field often has a high and variable level of measurement error, especially when different individuals take samples at different times.

9. See Chapter 5 for additional discussion of challenges associated with defining locations in mammal surveys.

10. Camera traps are cameras with motion sensors that take digital photographs when the motion detector is tripped.

11. As in the case of sample points within plots, these cameras within bands are pseudo-replicates. The sites themselves are the replicates. See also Chapter 9.

12. From these data, the probability that a mammal is present at a site ("occupancy") was estimated following methods in MacKenzie et al. (2002). This model includes estimators of detection probability. Additional details on how and why detection probability should be modeled within occupancy models is given in Chapter 15.

13. Recall the sampling of mammals and birds in the IBM. For mammals, the extent of similarity was defined based primarily on the nature of the rural village (for example, agricultural versus urban) or by the nature of the agriculture itself (livestock versus orchards). In contrast, vegetation configuration was considered more important for birds. As a result, the monitoring methodology for mammals and birds also differed: Mammals were sampled with passive camera traps, whereas birds were sampled by individuals using point counts.

## CHAPTER EIGHT

Creating Coherent Time Series through Repeated Measurements in a Marine Monitoring Program

Yonathan Shaked and Amatzia Genin

Environmental monitoring encompassing diverse methods and wide spatiotemporal ranges uses repeated measurements to form coherent time series. Because each sample is unique, repeated sampling is not expected to yield identical results. In this setting, "replication" has a context-dependent meaning that supports specific goals within each sampling campaign.

Environmental monitoring programs often are long-term, large-scale operations, initiated after environmental deterioration has been recognized, and implemented with the goal of informing decisions regarding current or future activities—be it the determination of acceptable levels of environmental impact or the application of restoration efforts. Regardless of when or how monitoring programs are established, however, they need to provide reliable data collected through time in repeatable ways. In this case study, we discuss how the design and implementation of Israel's National Monitoring Program (NMP) in the Gulf of Eilat explicitly incorporates concepts of replication and repeatability to ensure the collection of reliable data that can be used as a tool for effective environmental management.

RATIONALE FOR THE NMP

The NMP was created in 2003 to support decision makers in Israel's National Ministry of Environmental Protection (MEP) by "creating a long-term scientific database of the . . . oceanographic state at the northern Gulf of Eilat.[1] This database will be the scientific foundation for operational recommendations to address ecological problems, environmental management programs and continued informed development of the coastal region."[2]

The creation of the NMP came about in the aftermath of a long and heated public debate about industrial aquaculture in the northern Gulf of Aqaba off the shores of Eilat (fig. 8.1). Aquaculture in the Gulf of Aqaba began in the late 1980s with the idea of raising marine fish on land using water pumped from the sea, but the enterprise eventually evolved

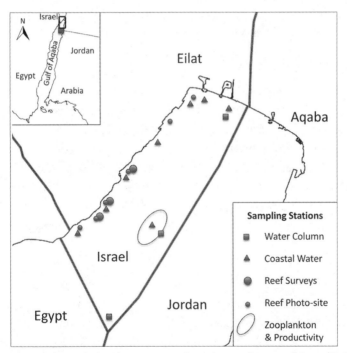

FIG. 8.1. Map of sampling and survey sites along the Israeli Coast of the Gulf of Aqaba. The map shows sites visited regularly by Israel's National Monitoring Program at the Gulf of Eilat. Inset at top left is a regional map, with the area given in detail marked by an open rectangle, and it shows the location of the southernmost oceanographic sampling station, out of Israeli waters.

into the more standard use of offshore cages in which fish are raised. This industry grew considerably throughout the 1990s and into the early 2000s despite growing worldwide awareness of the decline of coral reefs and reports of deterioration of the reefs of Eilat lining the shores close to the aquaculture pens.[3] Concerns were raised about possible detrimental effects of aquaculture in Eilat on the reefs. Counterclaims suggested that these effects either were negligible or even increased coral growth.

Both environmentalists and aquaculture's supporters looked to scientists for data to support their positions. However, not only were there insufficient data to determine the effects of aquaculture on the coral reefs and ecosystem dynamics in the Gulf of Aqaba but also the scientists themselves could not agree on the validity of the data.[4] The need for reliable and consistent monitoring of the Gulf became apparent, and the NMP was established for that purpose, to provide baseline data on the state of the Gulf and its coral reefs, and to yield time-series data that could inform environmental management decisions.[5] The last aquaculture cage was removed in the summer of 2008.

DESIGN AND IMPLEMENTATION OF THE NMP
At the heart of the NMP are two very different habitats, each with its own sampling and monitoring challenges (see fig. 8.1). One habitat is the water column, from the surface to a depth of approximately 800 meters. The other habitat is the shallow coastal region, including the coral reefs. In these two habitats, the NMP takes different sets of measurements using different procedures that are sampled at different time scales; replication and repeated measurements have different meanings in these two habitats.

Despite differences in the settings and habitats monitored by the NMP, however, common guidelines have been considered in its design and implementation (see also Chapter 6), which apply to all its activities. First, clear and coherent goals were defined in its mission statement. Second, indicators, variables, and appropriate time scales for monitoring the variables in each habitat were identified based on previous research and in consultation with numerous researchers studying the Gulf of Aqaba. Third, sampling methods and protocols reflected expertise, logistic constraints, and budgetary realities. Fourth, data are subject to quality control to ensure consistency and accuracy before processing or public release.[6] Finally, the NMP was designed with adaptive monitoring in mind, providing researchers with the ability to change parts of the monitoring program as progress was made so as to best serve its goals as new knowledge accrued.[7]

Continuous online instrumental measurements of ambient physical variables such as water temperature, sea level, and current velocity are taken at fixed locations and streamed to data loggers together with meteorological measurements. Instruments are calibrated and cross-checked semi-annually.

Sampling of the water column is divided into several campaigns, usually repeated at monthly intervals from a research vessel. At fixed locations (i.e., within ca. 100 m) based on global positioning system coordinates) samples are taken at fixed depths (depths are accurate to within 2 m) to produce a profile of the water column. Accumulation of these monthly profiles produces a time series (fig. 8.2). Constraining the sampling locations in the open sea to within 100 m of one another was deemed to be sufficient because seawater away from shore is thought to be laterally homogeneous over the kilometer scale. Sampling depth, however, was constrained more precisely because many chemical and physical properties change with depth.

Seawater samples are collected at each location and depth combination using 11-L Go-Flo bottles; each sample is divided on board the research vessel into numerous sub-samples for laboratory analysis of chemical and biological (plankton) variables. Analytical precision and reproducibility of analytical results is achieved by analyzing duplicate or triplicate sub-samples (analytical replicates) from a given seawater sample alongside laboratory standards. Additional cross-reference for some measurements (e.g., of salinity) is provided by electronic measurements taken at the same time that the water sample is collected.

Another monthly cruise samples the sea surface along several fixed near-shore locations. This is usually done within one to two days of the water-column cruise. A control sampling station over deep water is used for evaluation of near-shore results in the context of local pollution input from shore and is expected to yield similar (comparable but not necessarily identical; see Chapter 15) results to the samples from the sea surface

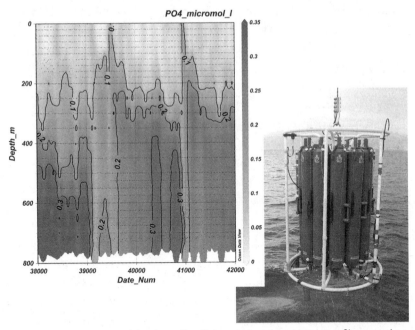

FIG. 8.2. Time-series and depth profile of phosphate concentration profiles near the joint Israel-Egypt-Jordan border. The water sampler is shown in the photograph (inset). At the location where these data were collected, the water depth is 730 m. Each black dot on the graph represents a water sample collected and analyzed. A vertical line of dots represents samples at different depths from a single cruise, creating a water column profile. Taken together, samples from regular National Monitoring Program monthly cruises done over the past decade show seasonal and inter-annual changes in phosphate concentrations. Plot created using ODV software. ODV software written by R. Schlitzer, Ocean Data View, 2015. Available at http://odv.awi.de. Inset photo by I. Agalon, and used with permission.

taken on the water-column cruise (fig. 8.3). Each month plankton tows are used to assess zooplankton biomass and primary productivity is measured using radiocarbon methods.[8]

Although samples from different sampling locations, depths, or times are not expected to be identical, for all measurements the NMP expects analytical replicates to be identical within instrumental precision limits. In addition, many variables have well-established correlations and relationships (e.g., depth profiles; see fig. 8.2) or exhibit seasonal variability within predictable limits. These limits add additional levels of quality control for measurements that cannot be repeated precisely.

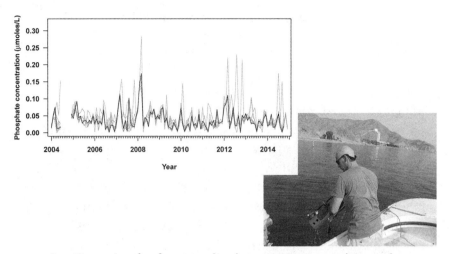

FIG. 8.3. Time series of surface water phosphate concentrations at the coastal stations visited monthly by the National Monitoring Program. The graph shows concentrations of phosphorus at the various stations (thin pale lines) plotted against that of the reference station (thicker dark line), and deviations are attributed to local input from shore. The inset photograph illustrates the sampling process. Photo by Nitzan Seger-Shanlov, and used with permission.

In the shallow coastal habitat, the monitoring is focused primarily on ecological conditions that are meant to assess and characterize key populations (see Chapter 6), such as corals, mobile nocturnal invertebrates and fish, and their habitats at different sites along the coast. Annual surveys are done at fixed sites (fixed for a combination of location and depth) by divers who identify and record the abundances and characteristics of target species and their habitats. At each site, random, haphazard, or structured sub-samples (replicates) are used to achieve the desired precision of statistical estimates (see also Chapters 9, 10). The sampling strategy and procedures are fixed across sites for each target population, but the number of samples collected is determined based on results of the actual on-going survey according to the natural variability encountered each year at a particular site and guided by the desired statistical precision.

A different approach to repeated sampling in time is done using annual (late summer) photographic surveys of fixed locations on the reef. Photographs are taken with a camera attached to the reef so that photos from different years largely overlap. Each photograph is digitized, and key features (e.g., individual coral colonies) are identified. Time series of photographs allow NMP researchers to follow the fate through time of particular colonies, providing data that complement results from the field surveys.

REPLICATION IN NMP TIME-SERIES DATA

As in the case of Israel's National Biodiversity Monitoring Program (Chapter 7), replication and repeatability were of paramount concern in the design of the NMP and are regularly evaluated during field sampling campaigns and laboratory analyses. Similarly to most monitoring programs (Chapter 6), different measurements represent unique points in time, and each subsequent sample is not expected to reproduce previous results. Rather, by using repeatable protocols and calibrated instruments, measurements of the same variables at the same locations are used to document environmental changes.

The NMP repeatedly measures and documents as many ecologically relevant variables as is logistically possible so as to reveal relationships and trends among them. With additional evidence for cause-and-effect relationships, these data can inform sound management decisions.[9] Although long-term monitoring provides a natural and powerful approach to studying complex natural systems (see Chapter 6), it can take many years before statistically significant trends emerge from the background noise of year-to-year variability.

Replication serves specific goals within different sampling campaigns, and so carries different, context-dependent, meanings. The primary aim of the NMP is to identify ecological trends and their underlying causes along the shores of Eilat. Whereas trends are determined directly from the data and are measured relative to a baseline, causes are rarely straightforward to identify, and measurements require context.[10] The need to provide context for data that have no "correct" values highlights the problem of replication. Each measurement is unique and cannot be repeated exactly to reproduce identically a previous one.[11] Rather, repeated measurements—taken using identical protocols, instruments, and staff—define trends and boundaries of variability over time. Individual measurements taken at the same place but at different times are not expected to be the same, because time has elapsed and the system has undergone chemical, biological, and physical changes. In the marine environment, differences between repeated measurements also are amplified by the mobility of the medium. Thus, not only has the water body evolved and changed since the previous measurement; the researchers are not even measuring the same waters.

DISCUSSION

Israel's national monitoring program in the Gulf of Eilat is focused on monitoring—repeated sampling through time—to create a time series of data that can identify environmental changes. Measurements taken by the NMP are meant not to be reproduced but rather to be repeated at different times at fixed locations (see also Chapters 5, 15, 16) so as to yield comparable results that document change.

By incorporating a variety of approaches and procedures, the NMP addresses replication on several levels (see also Chapter 2). Analytical replicates—repeated measurements of identical samples or standards— identify and constrain equipment errors and determine analytical precision. Environmental replicates, such as sub-samples of a site in ecological surveys or discrete water samples, are expected to reflect natural variability and to place bounds on the precision of statistical estimators. Having a dedicated technical staff comprising multiple individuals who collect survey data helps reduce individual biases in field data collection. Finally, repeated sampling campaigns are comparable (see Chapter 15) because the NMP uses the same sampling and analytical protocols year after year and in ways that account for variables (natural or otherwise) that are not the objects of monitoring.

For example, many variables measured by the NMP have daily or seasonal signals. Where identifying these rhythms is part of the monitoring program (e.g., by characterizing daily changes in plankton population sizes or seasonal changes in seawater chemical properties), sampling intervals are used that can detect the desired signals. In contrast, where intra-annual cycles (e.g., seasonal cycles of corals, algae, mobile invertebrates, or fish) are not part of an indicator variable, samples are taken annually at similar times of year.[12]

Controlling the locations or timing of sampling helps to identify signals of environmental change amid background variability and natural noise. But locations and timing can be controlled only within limits set by the context of the sampling campaign itself (see Chapter 16). An individual location—a reef site or a particular water-column sampling site—is defined broadly as a combination of geographical location, depth, or reef sector. Each sample is intended to characterize the specific location and to be comparable to samples taken at other times. Comparability, however, is defined by the goals of the sampling campaign.

Data from reef surveys are considered to be comparable when consecutive surveys are done during the summer (typically June to September); surveys done in January are thought to introduce unacceptable errors. For most monthly measurements, acceptable monthly intervals can range from 25 to 35 days, depending on weather and other logistical considerations. Similarly, when sampling the water column, the research vessel is likely to drift depending on currents and wind, inaccuracies are introduced into repeated locations, and a particular location on the ocean surface is defined with only 100-m precision. In coastal waters, greater precision in sampling location is required than in open water because monitoring changes in water properties along the coast are part of the survey's goals (see Chapter 16). In summary, the NMP replication takes on a range of meanings that are defined by the targeted goals of each campaign based on scientific judgment and outlined in the sampling protocols.

NOTES

The NMP team includes I. Ayalon, M. Chernichovsky, M. Drey, I. Kolesnikov, T. Rivlin, and N. Segev-Shaulov. The authors and the NMP team thank Israel's Ministry of Environmental Protection (MEP) for supporting the program and the Interuniversity Institute for Marine Sciences in Eilat for hosting it.

1. Long-term monitoring programs generally are not funded in the same way as short-term, academic research programs. The NMP is funded and supported by the MEP itself, but it is run by the Hebrew University and operated by the Interuniversity Institute for Marine Sciences in Eilat. The involvement of these long-standing, nationally supported institutes that have supported decades of research at the Gulf of Eilat has brought essential stability to the NMP. The NMP is governed by a steering committee consisting of the scientific director and representatives of the marine section at MEP, relevant members of academia, representatives of national bodies charged with nature preservation, and others involved in marine research in Israel. The NMP submits its annual scientific report to this committee, and the report is subject to its approval before being made public through an online database. In addition, the NMP or its steering committee may convene meetings of specific expert committees to help direct and further develop the

NMP. The NMP is staffed with specialists with overlapping skills. Most operations are joint efforts, which helps ensure continuity and also mitigates the individual biases that are particularly common in ecological surveys.

2. Israel National Monitoring Program at the Gulf of Eilat Annual Scientific Report (2014: 4). Translated from Hebrew by the authors.

3. See, for example, Angel et al. (2000); Pandolfi et al. (2003): Loya et al. (2004a, 2004b); Bruno and Selig (2007); Hoegh-Guldberg et al. (2007).

4. See Wielgus (2003); Loya (2007); and references therein.

5. Parallel efforts in other coral reef and marine ecosystems include those of Karl and Lukas (1996); Steinberg et al. (2001); and the nascent Ocean Observatory Initiative (OOI: http://oceanobservatories.org/).

6. NMP data are open to public use through the Web site of the Interuniversity Institute for Marine Sciences in Eilat (http://iui-eilat.ac.il).

7. See Lindenmayer and Likens (2009) and Chapter 6 for additional discussion of adaptive monitoring.

8. Both plankton composition and primary productivity are biological properties with inherent spatial and temporal variability. This variability, along with associated analytical concerns, is reflected in the sampling design and interpretation of replicability. Three consecutive replicate plankton-net tows are taken, each consisting of two closely coupled nets. Material from the coupled nets is pooled to yield enough material for analysis. The three consecutive tows are statistical replicates to estimate natural variability. To measure primary productivity, the upper 100 m of the water column is sampled at regular intervals. From each sample four subsamples are taken, three of which serve as replicates and one as a control sample. After treatment the sample bottles are returned to their respective depths for 24 hours before recollection for laboratory analysis.

9. See additional discussion in Beever (2006).

10. Determining the baseline is critical, as it defines the starting point against which any trend will be evaluated and provides context for the measurements. Measurements themselves, however, do not answer questions such as these: Is 20 percent average live stony coral cover on Eilat's coral reefs good or bad? Or is an average particular density of corallivorous *Drupella cornus* snails per colony of *Acropora* species a sustainable steady state or an ecological disaster? Of course, different coral reefs differ considerably from one another. A review and meta-analysis of previously published data from more than 300 reef sites in the Indo-Pacific region, ranging in coral cover from almost none to more than 100 percent, estimated the average live coral cover for the entire region to be 22.1 percent (Bruno and Selig 2007). Coral cover at the eight NMP-surveyed fore-reef sites in Eilat ranged from 12.2 percent to 54.9 percent in 2013, and the average cover calculated from these sites increased from 19.1 percent in 2004 to 26.4 percent in 2013. At the same time, a recent study in Eilat demonstrated that while rising *D. cornus* density temporarily lowered coral cover on the reef flat by eating specific

fast-growing corals, it helped sustain a high level of coral diversity (A. Zvu-loni, unpublished data). In short, measurements require context, and the NMP provides context by establishing and expanding the baseline of natural variability in ecologically relevant variables and checking its results against previous measurements and measurements of related variables. See also note 5 in Chapter 6 for a discussion of changing perceptions of baselines.

11. In the NMP, as is probably the case in most monitoring programs, reproduc-ibility of results is restricted to duplicates/triplicates. The sub-samples are used to check laboratory analytical procedures and to check for equipment errors.

12. See Chapter 16 for detailed discussion of variability and cycles in space and time.

Replication and Experiments

## CHAPTER NINE

Contingent Repeatability of Experiments in Time and Space

Aaron M. Ellison

Outcomes of biological experiments are contingent on when and where the experiments are executed. Concluding that an experiment has been repeated and its results reproduced is conditional on appropriate statistical analysis and accurate statistical inference.

Scientists are skeptics. Not skeptics in the original sense of the word—those ancient Greek followers of Pyrrho (360–270 BCE), who doubted the possibility of any real knowledge—but rather seekers of truth who continually question received wisdom and test others' claims of knowledge carefully and systematically.[1] In particular, scientific knowledge reflects our understanding of how the world around us "works," without resort to miracles, supernatural beings, or other unique and irreproducible phenomena. For a scientist, knowing how something "works" is equivalent to understanding cause-and-effect relationships. Applying the scientific method allows skeptical scientists to distinguish true causes from spurious ones and eliminate pseudo-causes or the occurrence of random events.

The scientific method is a deceptively simple recipe.[2] First, construct a range of different, ideally mutually exclusive, hypotheses: proposed explanations for an observed phenomenon based on first principles (e.g., mathematical or physical axioms or theories) or other available information derived from observations or previous experiments. Then, test the different explanations, using observations, experiments, and/or models. Note that the best test will falsify (*not prove*) one or more of these hypotheses. Repeat. After several rounds of this process of observation → hypothesis generation → hypothesis testing → hypothesis rejection, only one hypothesis will remain standing. Superficially, the scientific method resembles the deductive method (and maxim) of Arthur Conan Doyle's famous detective, Sherlock Holmes: "When you have eliminated the impossible, whatever remains, however improbable, must be the truth."[3] But unlike detectives and courts of law, for whom or which "beyond a reasonable doubt" is sufficient to convict, scientists should be ever skeptical of the "truth" and persist in trying to falsify even their seemingly most bullet-proof hypotheses.[4]

In short, skeptical scientists are always trying to disprove their pet hypotheses, both by making new observations and with experiments. Observations from monitoring programs, surveys, and structured samples (Chapters 4–8) regularly yield valuable data that can guide the development of scientific hypotheses, but such observations can provide evidence only of association, not causation.[5] The common maxim that *correlation is not causation* reflects the long litany of hypotheses that were supported by observations but crumbled under more intensive scrutiny.[6] In fact, scientific understanding advances most rapidly when existing explanations for observed phenomena are found wanting and new explanations are proposed and rigorously tested.[7] And, most importantly, scientists often learn the most about the systems they study when they fail to reproduce an experimental finding.[8] The inability to reproduce the results of an experiment may reveal the importance of precise methods. It also may point to unappreciated complexity and interactions among, for example, genes, species, or organisms and their environments and may eventually lead to new developments in theory.

It is important to distinguish between hypotheses, theories, and models, especially when discussing experiments and reproducibility. Hypotheses are proposed explanations for a particular, usually tightly defined, phenomenon.[9] Hypotheses are routinely tested by applications of the scientific method. Theories integrate results of observations and experiments into more encompassing explanations for a broad range of phenomena. Unlike hypotheses, theories are used to make predictions (forecasts) about unobserved phenomena in many settings.[10] Finally, models are simplified explanations for particular phenomena or systems under study. Models may be conceptual, graphical, or mathematical, but they always emphasize the "most important" parts of the system and de-emphasize, or omit entirely, other parts. As idealized representations of reality, models allow us to home in on particular factors to study, but because they do not include every possible cause or effect of interest, models can never be 100 percent correct.[11]

In spite of the many challenges arising from the simplification of systems into models and the experimental testing of one (or, at best, a few) causal factor(s) at a time (see also Chapters 10–12 and 17), experiments remain the central tool (or "gold standard") used by all scientists to identify cause-and-effect relationships and to separate true causes from false ones. Unlike in observational studies, which we do without nominally disturbing the system,[12] when we execute an experiment, we deliberately change the system.[13] In particular, we start with a hypothesis about how one or more variables, singly or interactively, cause measurable changes in a response variable of interest. We then deliberately alter the levels of the hypothesized causal variables and measure the effects of the manipulation. When experiments are carefully designed and adequately replicated (discussed briefly below and in more detail in Chapter 17), they can provide strong evidence for cause-and-effect relationships.

This simple procedure—manipulate the putative causes, measure their effects—is always conditioned on the assumption of *ceteris paribus* (all other things being equal). But are they ever, in fact, equal?[14] We regularly assume that if we work in a laboratory with, for example, standard reagents, well-documented protocols, and genetically identical mice, the results will be the same whether the experiment is done on a Monday or a Tuesday, in 2009 or 2012, in New York or Tel Aviv. But this is rarely true. For example, cutting-edge experiments exploring how genetic variability affects gene and protein expression or disease etiology routinely suffer from "batch effects," and their results are substantially affected by laboratory conditions (e.g., temperature, humidity), reagent lots, which individual organisms were tested, and who actually prepared and ran the samples.[15] *Ceteris paribus* is even harder to assert in ecological and environmental field research. Each day is obviously different, and a forest in Minnesota is not the same as a forest in Wisconsin. We rarely are surprised to get somewhat different results from, for example, an experiment examining how forests respond to the loss of their dominant trees in Massachusetts and in Pennsylvania or even from one done in the same forest stand in Massachusetts but in two different years.[16] In short, accurately identifying and specifying the same place or the same time is as difficult for laboratory research as it is for field research.

REPLICATION AND REPRODUCIBILITY

It is important to distinguish an experiment with replicates from an experiment that has been replicated—i.e., repeated (see Chapter 1). The growing literature on the difficulty of repeating research findings in the biological and psychological sciences often uses "replication" and "repeatability" interchangeably.[17] But, as we discussed in Chapter 1, "replication" by itself is a generic term that includes, among other concepts, both experimental replicates and experiments that are repeated independently. Thus, for clarity and consistency with existing literature on experimental design, we refer to the former as a "replicated experiment" and the latter as a "repeated experiment." If the results of a repeated experiment confirm the initial findings, we would consider the experimental results to have been "reproduced" successfully.[18]

At the same time, the importance of appropriate replication within experiments cannot be overstated. In both field and laboratory settings, researchers expect that there will be stochastic or unmeasurable differences among individuals—even among lab-mice littermates or clones. Thus, building adequate replication—and using truly independent replicates—into an experimental design is necessary to separate actual causal effects on an outcome of interest from small but observable differences associated with variability among individual replicates. Subsequently, analysis of experimental data using classical statistical techniques such as the analysis of variance (ANOVA) can be used to examine and test the difference between means or variances among experimental groups (i.e., groups of replicate individuals subject to different manipulations) from variance within experimental groups (i.e., all the individuals subject to the same manipulation).[19]

By how much do individuals in one experimental (treatment) group need to differ from individuals in another treatment group for the observed differences between groups to be considered "significantly" different? Conversely, how similar is "close enough" for them to be considered "the same"? Unfortunately, the answer is neither simple nor straightforward, and it involves four related quantities: sample size, effect size, the P value, and statistical power. These principles apply whether the research is a field study with few (e.g., < 50) replicates or a genome-wide association study looking at thousands or tens of thousands of genes.

Sample size refers both to the number of individuals within *each* treatment group and to the total number of individuals added up among *all* treatment groups. But the former is much more important than the latter, because as the number of individuals within a treatment group increases, the precision of the estimate of variance within the group (an important measure of variability) also increases.[20]

The variance of a sample of measured variable (e.g., the binding efficiency of a protein or the height of a tree) of a group of measurements is estimated as $s^2 = \frac{1}{n}\sum_{i=1}^{n}(y_i - \bar{y})^2$, where y_i is the value of the measured variable for the ith individual replicate and $\bar{y}$ is the observed average (or mean) value of the measured variable for all n individuals measured $\bar{y} = \frac{1}{n}\sum_{i=1}^{n}y_i$, where $\sum_{i=1}^{n}y_i$ means "add up" all of the n observations: $\sum_{i=1}^{n}y_i = y_1 + y_2 + \ldots + y_n$.[21] In the case of the variance, the quantity added up is the squared differences from the mean: $\sum_{i=1}^{n}(y_i - \bar{y})^2 = (y_1 - \bar{y})^2 + (y_2 - \bar{y})^2 + \ldots + (y_n - \bar{y})^2$. If there is only one replicate, there is no variance, as the "average" value is simply the value measured for a single individual, and the difference of the observation from the average (itself) equals zero. Because the variance cannot be calculated without replication, data from unreplicated experiments (i.e., experiments without individual replicates) cannot be used to test for differences between populations. When results of studies are combined in a meta-analysis, the key quantities that are needed are the mean response and its associated standard deviation (see Chapters 11, 12). The estimated standard deviation s is simply the square root of the variance ($s = \sqrt{s^2}$). So if the results of an experimental (or observational) study do not have any variance because it was unreplicated, those results also cannot be used in a synthetic meta-analysis.

In short, using 10 individuals spread among 10 treatment groups (including a control) does not allow one to estimate the variance *within* each treatment group. This lack of within-group variation results in much less statistical power to detect differences among groups than would, for example, an experiment with 10 individuals assigned randomly to two treatment groups.[22] Nonetheless, the literature is replete with analyses of unreplicated experiments. Examples from ecology include the simulated hurricane and nitrogen saturation experiments at Harvard Forest, among many others.[23] In these and many other unreplicated experiments, researchers often sub-sample the unreplicated plots and inappropriately treat the sub-samples as replicates. It should therefore not be surprising that most attempts to reproduce unreplicated experiments yield very different results.[24]

But it is not enough simply to have experimental replicates. Unbiased estimation of variance also demands that the replicates be independent of one another.[25] The essential difficulty with pseudo-replicates (e.g., sub-samples from the same treatment plot, replicate blood samples from a single individual) is that they are not independent. They may be so close to one another in space or time that they actually affect each other in unmeasured ways, or the applied treatment may not have actually been applied uniformly over the plot. The latter problem is exacerbated further with large-scale experiments. Lack of independence can arise in many other ways, and it is incumbent on the experimenter to present a convincing case that replicates are truly independent.

How many replicates are needed? At a minimum, only two replicates are needed to estimate their variance, but the accuracy and precision of the estimate of the variance, like the accuracy and precision of the estimate of the population mean, increases substantially with sample size (fig. 9.1). For example, the average height of an adult male in the United States is approximately 175 cm, and the standard deviation is 11.5 cm (as the standard deviation is simply the square root of the variance, the variance $\approx$ 130 cm²).[26]

The relationship between the sample size, accuracy, and precision of an estimator is independent of whether the data come from an observational study or an experimental manipulation (see also endnote 20

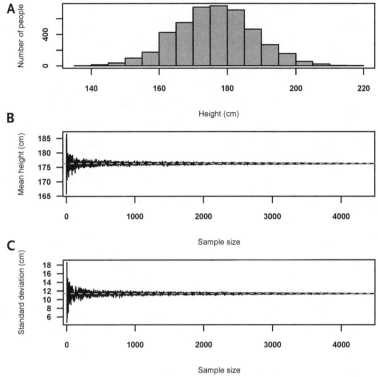

FIG. 9.1. Illustration of the importance of sample size for the mean and variance. The plots are based on a simulation of a sample of 4,482 adult male heights (A) with mean = 176.3 cm and standard deviation = 11.38 cm (based on data in note 27). As the number of observations drawn from the sample increases, the estimates of the population mean (B) and standard deviation (C) from the sample converge on the "true" value (dashed gray line). Note that all of the estimates are accurate and unbiased, but the precision increases dramatically with sample size.

and Chapters 1 and 15). But in an experimental manipulation, we are usually interested in comparing parameter estimates among two or more groups.[27] An ANOVA uses differences in estimated variance within and among treatment groups to help determine whether the measured differences in, for example, means among groups are meaningful. If, as suggested (see fig. 9.1), the estimate of mean or variance in height varies widely with fewer than 50 observations, it should come as no surprise that experiments with far fewer replicates nonetheless yield "significant" results that are difficult to reproduce successfully.

HOW CLOSE IS CLOSE ENOUGH?

How similar do two repeated runs of a given experiment have to be for us to consider that the results of the initial study have been confirmed successfully? Answering this question requires some understanding of P values and statistical power. As noted above (and see endnote 9), we are most familiar with P values and the idea that the result of a study is "statistically significant" when $P \leq 0.05$. What this really means, however, is that the probability (P) of falsely rejecting a true null hypothesis (that there is no cause-and-effect relationship between our hypothesized cause and an observed effect) is less than or equal to 5 percent. In other words, as skeptical scientists, we turn our pet hypothesis (A causes B) on its head to create a null hypothesis (A and B are unrelated), then use our data to test the null hypothesis. If we reject the null hypothesis (with a probability of 5 percent or less), there is some evidence that, just maybe, our pet hypothesis is true.

But it turns out that rejecting the null hypothesis with a probability that we made a mistake equal to 5 percent is not equivalent to accepting the alternative hypothesis with a 95 percent certainty. Rather, there are two other alternative outcomes to our experiment in this simple universe (table 9.1). Our data could lead us to correctly accept our null hypothesis, incorrectly accept our pet hypothesis, or correctly accept it. In this four-part probability matrix, there are two "correct" decisions based on our data, and two errors, which we call Type I and Type II statistical errors.[28]

Since we have four possible outcomes, the sum of their probabilities has to equal one, so the probabilities of any two must be less than one; by definition, therefore, the probability of correctly rejecting a false null

Table 9.1

The four-fold path of statistical hypothesis testing

| | | EXPERIMENTAL RESULTS SUGGEST: | |
		Accept H_0	*Reject H_0*
"Truth"	H_0 is true	Correct decision	Type I error (a)
	H_0 is false	Type II error (b)	Correct decision

Note: H_0 is the null hypothesis of no relationship between causal factor and observed outcomes. The value that each decision can take on is a probability, so each can range between zero and one. Statistical power equals one minus the probability of a Type II error.

hypothesis plus the probability of incorrectly rejecting it must be less than one (100 percent); statistical power equals one minus the probability of committing a Type II error.

Through careful experimental design (see Chapter 17) and sufficient within-experiment replication, we can minimize the probability of committing both Type I and Type II errors and also minimize the probability of making the "wrong" correct decision, but we can only get close to certainty, not all the way there.

Furthermore, the more studies we do, the more likely it is that we will get at least one "significant" effect due to small differences in methods, people, time of day, or random chance alone. Thus, when we look at results from two or more "identical" (or at least repeated) experiments, we need to adjust our cut-off for statistical significance to account for the possibility that we will make a Type I error by chance alone. The two general classes of methods used to control for errors in multiple hypothesis testing are called "family-wise error rate" methods and "false discovery rate" methods. Family-wise methods, such as Bonferroni corrections, are very stringent and are designed to reduce the probability of even a single Type I error among all the repeated experiments. False discovery rate methods aim to reduce the expected proportion of Type I errors among the repeated experiments.[29]

Both family-wise and false discovery rate methods are themselves probabilistic, so there can never be 100 percent certainty that an experiment was repeated and its results were reproduced exactly once or replicated multiple times. Quantification of these probabilities is especially important in large-scale experiments involving gene expression or various -omes (proteomes, transcriptomes, etc.) and is an area of active research.[30] What we can be certain of, however, is that with more careful attention to experimental design and detailed explication of methods (summarized with recommendations in Chapter 17), we are more likely to minimize errors of inference based on our results. This conclusion is illustrated in the next case study (see Chapter 10).

NOTES

1. "Skeptic" is derived from the Latin *scepticus*, meaning inquiring and reflective, and was applied to the disciples of Pyrrhos. By the Renaissance, it had come to mean one who was inclined to disbelief, particularly with respect to

metaphysics, spirits, and Scripture. See Agassi and Meidan (2008) for its relevance to present-day science and philosophy and the *Oxford English Dictionary* online (http://www.oed.com/), accessed 13 March 2014.

2. See Chamberlin (1890); Platt (1964); and Taper and Lele (2004) for readable overviews.

3. *The Sign of Four* (Conan Doyle 1890: p. 111).

4. The foundational text on the hypothetico-deductive, aka scientific, method is Popper (1959). Feyerabend (1975) provides a philosophical critique of Popper, but his approach has not been adopted widely by the scientific community. Bayesian inference (after Bayes 1763) generally is viewed as an alternative inductive and confirmatory approach to evaluating scientific hypothesis testing (e.g., Savage 1954; Howson and Urbach 1989) and one that is applied more readily to observational data than to experimental data. For a contrary view that places Bayesian inference firmly in the hypothetico-deductive realm, see Gelman and Shalizi (2013).

5. Well-designed observational studies and monitoring programs often yield results that are qualitatively similar to experimental findings (e.g., Elmendorf et al. 2015; see Chapter 8). Nevertheless, the former much more rarely provide strong evidence for cause-and-effect relationships.

6. A contemporary example is provided by the hypothesized relationship between early-childhood vaccination and the development of autism (reviewed by Doja and Roberts 2006). The initial correlation was based on small sample sizes (12, then 60) and failure to control for co-occurring conditions or financial conflicts of interest. Subsequent large-scale observational and longitudinal studies (involving hundreds of thousands of individuals) followed by a matched case-control study (with thousands of individuals) found neither a correlation nor a causal relationship between vaccines and autism.

7. As we discussed in Chapter 1, the decline in the use of experiments in physics is considered a threat to the intellectual integrity of the discipline (Ellis and Silk 2014).

8. The process of explicitly learning from failure is widely appreciated not only in the biological and physical sciences (e.g., Livio 2014) but also in engineering (e.g., Petroski 1985) and software development (e.g., Neumann 1994).

9. It is normal practice for experimental biologists to formally state a "null hypothesis" (Ho: there is no relationship between a suspected cause for an observed effect; for example, Ho: there is no effect of vaccines for rubella on the development of autism). The experimental data are then used to "test" this null hypothesis using statistics. The result is a probability (or P) value that expresses the probability of incorrectly rejecting this null hypothesis. A low value, such as $P < 0.05$, suggests that it is unlikely that the null hypothesis was falsely rejected. However, a P value equal to 0.05 does not mean that the alternative hypothesis (and usually there is more than one) would be accepted with $1-P = 0.95$ (or 95 percent) probability. See Ellison

(1996) for a more detailed discussion of this issue and the Bayesian alternatives to *P* values.

10. "Evolution by natural selection" and the "formation of the universe following a singular event" (the "Big Bang") are two examples of theories. Each
was derived from many observations and experimental tests and has been
found to make accurate predictions in many disparate contexts. Theories are
broadly accepted and rarely tested (see Kuhn 1962 for additional discussion
of the consequences of critically testing scientific theories).

11. The statisticians George E. P. Box and Norman R. Draper famously wrote
(1987: 424) that "all models are wrong, but some are useful." The evolutionary biologist Richard Levins, reflecting on the role of models and theories in
evolutionary biology, wrote (1966: 20), "If these models, despite their different assumptions, lead to similar results, we have what we can call a robust
theorem [i.e., a theory] that is relatively free of the details of the model.
Hence, our truth is at the intersection of independent lies." The key points
here are that models, being simplifications of reality and laden with assumptions, are not "true" but that if multiple independent models (i.e., ones with
different assumptions, different starting points, etc.; see Wimsatt 1980 for a
critique) all give the same results, an explanatory theory that builds on these
models is a reasonable predictive framework for reality.

12. Nonetheless, many so-called observational or monitoring studies have unintended or unanticipated effects as great as those of some manipulative experiments. Obvious examples include soil compaction from repeatedly
walking the same paths to reach a sampling station; perforation of soil
through repeated removal of soil core samples (aka the "Swiss cheese effect"); and behavioral alterations in small mammals that become trap-happy
or trap-shy following repeated bouts of trapping. But more subtle and unexpected effects of monitoring and data collection also have been reported. For
example, plant growth and herbivore impacts change in response to repeated
non-destructive measurements (Cahill et al. 2002). In a large-scale free-air
carbon dioxide enrichment experiment, trees from which regular observations of tree diameter, height, and canopy light transmission were taken had
up to 50 percent less biomass than unmeasured trees in the same plot (De
Boeck et al. 2008). Still other effects may be mediated through organisms
that are not the primary focus of the study. For example, disturbance associated with the placement of nutrient-capturing resins in soil may create
hotspots of activity of soil microarthropods, leading to localized changes in
nutrient availability unrelated to experimental manipulations.

Of course, all observations and sampling methods will impact the system to some extent (see also Schrödinger 1935). If these effects are minor
and are comparable among observational units or experimental treatments,
they are not a serious problem. But if the monitoring effects are strong or if
there is an interaction between monitoring activity and treatment (e.g.,
treatment plots are more trampled than control plots), the observational or

experimental results may be compromised and the conclusions derived from them may be unreliable.

13. There is a long history in the ecological literature of distinguishing "natural" or "mensurative" experiments from "manipulative" experiments. We consider "natural" and "mensurative" experiments to be equivalent to observations and "manipulative" experiments to be equivalent to experiments. Diamond (1986) discusses different experimental types in ecology, and DiNardo (2008) discusses their application to economics and the social sciences.

14. On philosophical problems with ceteris paribus and a defense of natural "laws" that are not philosophically defensible as ceteris paribus, see Cartwright (1989, 2002).

15. Leek et al. (2010) provide a current review of batch effects in biology laboratories. Remarkably, both the amount of gene expression observed and a substantial percentage of imputed interactions among genes change from batch to batch.

16. Compare Yorks et al. (2003) with Orwig et al. (2013) or Sullivan and Ellison (2006) with Farnsworth et al. (2012).

17. A good entrée into this literature is the collection put together by the editors of *Nature:* http://www.nature.com/nature/focus/reproducibility/.

18. This is a somewhat looser usage of "reproduced" than was suggested in Chapter 1. In that chapter, we considered a study to be reproduced if the investigator was able to get the exact same results after repeating the analysis using the same data at a different time yet without an additional process of empirical measurement. The latter is analogous to re-running computer code on a data set at a different time.

19. Standard references on experimental design and analysis using ANOVA include Fisher (1925); McCullagh and Nelder (1989); Underwood (1997).

20. The meaning of, and difference between, accuracy and precision is quite important. Accuracy is the degree to which the value of a measurement of an object comes close to the actual, true value. Precision is the level of agreement among a set of measurements on the same individual. For example, if a 1.75-m-tall person is measured in his physician's office to be 1.6 m tall, we would say that the measurement is not very accurate. But if this person was measured ten times and the difference between each measurement was only 1 mm, we would say that the measurements were quite precise.

21. We use Roman letters (s^2) for our estimate of the sample variance to distinguish it from the underlying (and unobserved) "true" value of the variance of the population. The latter we write with a Greek letter:

$\sigma^2 = \dfrac{1}{n}\Sigma_{i=1}^{n}(y_i - \mu)^2$, where μ is the population mean (also unobserved). For estimated parameters other than the familiar mean ($\bar{y}$), standard deviation (s), and variance (s^2), statisticians usually indicate parameter estimates with "hats": $\hat{\sigma}^2$ is the estimate of the population variance, and s^2 is an (asymptotically unbiased) estimate of $\hat{\sigma}^2$.

22. Strictly speaking, this applies only to ANOVA designs where the treatment groups are discrete variables, such as "children" and "adults." If the applied treatment, for example, antibiotic dosage or atmospheric CO_2, is a continuous variable, the data will be analyzed better using linear or non-linear regression.

23. Cooper-Ellis et al. (1999); Magill et al. (2004). Carpenter (1989, 1990) discusses the costs and benefits of unreplicated experiments when dealing with large ecological systems (e.g., whole lakes).

24. Hurlbert (1984) provides a detailed discussion of such "pseudo-replication" in biological research.

25. More precisely, this second requirement is that the individual replicates be independent and identically distributed. Not only should the replicates be independent from one another, but also each should have the same underlying probability distribution. See Gotelli and Ellison (2012) for additional discussion.

26. The actual value for all US males older than 20 years is 176.3 cm, with a standard error of the mean equal to 0.17 cm. The sample size for this estimate was 4,482. See McDowell et al. (2008), which reports the standard error of the mean: 0.17 cm. The standard error of the mean is an estimate of the standard deviation of the distribution of sample means of identical sample sizes and is calculated as the sample standard deviation divided by the square root of the sample size: $SE = \dfrac{SD}{\sqrt{n}}$, so the sample standard deviation of the estimate by McDowell et al. (2008) of male heights is $SD = 0.17 \times \sqrt{4482} = 11.38 \text{ cm}$. The estimation of the standard error of the mean is based on the observation that if a different 4,482 individuals were sampled, the sample mean would be different from that of the first sample. If this were repeated many times, a distribution of sample means would be obtained. The standard error is an estimate of the standard deviation of this distribution of sample means. For large sample sizes (>100 or so), a 95 percent confidence interval can be approximated as $\bar{Y} \pm 1.96 \times SE$. See Gotelli and Ellison (2012) for further discussion.

27. Many observational studies are not focused on comparisons among groups, but estimates of variance are still needed both to describe the variability among the observations and for use in subsequent meta-analyses or prospective power analyses, which are useful for future studies.

28. Type I errors are often called "producer errors" because a manufacturer assumes that there is no problem with a product until it is proven otherwise. Hence, very strong evidence is required to reject the null hypothesis of no problem, which might lead to a recall of the produce or its withdrawal from market. In contrast, Type II errors are often called "consumer errors" because consumers assume that a product is harmful until it is proven

harmless. Consumers want producers to minimize Type II errors, whereas producers want consumers to trust that there is a low level of Type I errors. This conflict lies at the heart of consumer product legislation and the precautionary principle. See Sokal and Rohlf (1995); Andorno (2004).

29. See Abdi (2007) for a review of family-wise error rate methods and Benjamini and Hochberg (1995) for an introduction to false discovery rate methods.

30. Heller et al. (2014) introduced this probabilistic statistical measure for replication in genome-wide association studies that minimize the impacts of false positives. Their measure can be applied either to family-wise error rate or false discovery rate methods.

## CHAPTER TEN

The Influence of Variation among Replicates on Repeatability

Jacob Pitcovski, Ehud Shahar, and Avigdor Cahaner

The distribution of individual replicates within a group can be as important as differences among group means, and it affects our ability to repeat an experimental result. When within-group variability can be accounted for in models that include measured covariates, repeatability of treatment means may be improved.

"Replication" is treated differently in industry, agriculture, medicine, and scientific research. For example, elucidating the sequence of events that leads to antibody production following vaccination requires replicates within experimental treatments (see also Chapters 1 and 9). In contrast, evaluating the therapeutic aim(s) of vaccination (protection against a given pathogen; see Chapters 11, 12) requires repetition of entire experiments.

Through discussion and analysis of two case studies drawn from experimental immunology, we first illustrate how a large number of observations can inform the design of a particular experiment intended to test rigorously the causes of the observed associations (see also Chapter 9). We then show that it is important when repeating experiments to reproduce not only the expected result (i.e., the estimated mean value) but also to reduce the variability in estimates of the mean. Reduction in overall variance can be accomplished both with more precise measurements and by accounting for additional sources of error (covariates or "nuisance" variables). When potentially important covariates are recorded during an experiment, they can be included in the analysis of the data and help to isolate the true "signal" of the experimental treatment from the "noise" of the experimental environment.

IMMUNE RESPONSES: MORE THAN THE MEAN

The receptors of the major histocompatibility complex (MHC, or HLA in humans) influence various immune characteristics. Each individual has his or her own "personal haplotype": the cell membrane of every one of our cells (excluding red blood cells) presents hundreds to thousands of copies of 3–6 different HLA molecules out of the range of thousands of HLA molecules that exist throughout the human population. Individuals with different HLA haplotypes respond differently to infection and vaccination.

Most people do not know their personal haplotype, nor do they know how their immune system responds at the cellular level to vaccination (the antibody "titer" or the quantity of antibodies produced in response to a vaccine). Consequently, individuals rarely know if or to what extent they actually are protected against a pathogen after being vaccinated.

Researchers normally study immune responses of entire populations, not of specific individuals.[1] Haplotypes differ among both individuals and subgroups ("populations"), such as Icelanders, Inuit, Ashkenazi Jews, etc., and thus there also are differences in immune responses to pathogens among these populations. Because antibody levels produced in response to a vaccine also may depend on personal haplotypes, different sub-populations may exhibit similar antibody levels on average, but the distribution of antibody levels among individuals can still differ among the populations.

Indeed, such differences are observed for hepatitis B; in most populations, about 5 percent of the individuals have particular HLA haplotypes that do not produce antibodies after being vaccinated against the pathogen.[2] Two populations may exhibit similar antibody averages yet important differences (e.g., 4 percent versus 6 percent) in the proportion of "non-responsive" individuals: those individuals that do not respond to vaccination because of genetic (e.g., HLA haplotype) or other reasons (e.g., immunosuppression). Thus, to repeatedly identify non-protected individuals—those with insufficient levels of antibodies or no antibodies at all—it is more important to assess within-population variability than the more common repeatability of the mean.

Similarly, the variation in antibody levels among individuals within a population (typically also represented by a random sample) still can be quite large and usually depends on sample size (see fig. 9.1 in Chapter 9). Such variation is commonly expressed by the sample variance (s^2) or sample standard deviation (s or SD). In any experiment, it is desirable that the SD be as small as possible. Repeated experiments often strive to further reduce the SD so as to gain further confidence that the estimate of the population mean is reliable.

The repeatability of experimental results is associated with statistical significance (see also Chapter 9). Studies usually are considered to be repeatable if each repetition yields the same mean response (see Chapters 11, 12); deviations from expected means (i.e., variation among replicates within or among treatment groups) may be dismissed as experimental errors or mistakes in techniques. However, if two (or more) groups repeatedly subject to similar treatments yield similar means but very different levels of variation, these groups may differ in important ways. Thus, the significance and repeatability of differences in variation among replicates also are important to consider.

CASE STUDY 1: EFFICACY OF VACCINES FOR NEWCASTLE DISEASE VIRUS IN POULTRY

Infectious diseases in poultry production are prevented by vaccination. Chicks are sufficiently protected by maternal antibodies only for the first 10–14 days of life. After that, maternal antibodies are significantly reduced in abundance and chicks become susceptible to a variety of infections.[3] Continuous protection from birth until marketing age (5–7 weeks) of chicks reared for meat ("broilers") is achieved by vaccination with attenuated or inactivated virus to stimulate the chick's own production of antibodies. However, high levels of maternal antibodies reduce the efficacy of vaccination at one day of age. The gap between the reduction in protective maternal antibodies and the elevation of self-produced antibodies following active vaccination is the period when chicks are most sensitive to pathogen infections.[4]

Newcastle disease virus (NDV) is one of the most prevalent infectious diseases of poultry, and it is controlled by vaccination. However, it is difficult to predict the outcome of vaccination in each population. The aim of this study was to identify both the mean titer of antibodies produced against NDV and the SD of the mean in commercial flocks of broilers. The observations were used to design a subsequent controlled experiment of the efficacy of the NDV vaccine.

It is normally expected that a uniform vaccination protocol that uses the same antigen source, route of vaccination, age of birds, and rearing conditions will yield identical production of antibodies and protection against a specific pathogen. As a result, samples of chicks taken at random from commercial flocks that have been vaccinated should repeatedly yield the same mean antibody titer. The key question here is whether the distribution of antibody production is also the same within each population.

Chicks with low titers (1 and 2) are considered to have insufficient levels of protective antibody; only chicks with titer ≥ 3 are considered to be sufficiently protected from NDV.[5] Thus, for assessing the protection of a flock, the distribution of individuals with low titer values among a tested sample of chicks is more critical than the mean titer of any given sample.

The data for this study consisted of NDV antibody titers in 3126 samples.[6] Each sample consisted of 23 chicks of the same age. The samples were taken at random in 2012, 2013, and 2014 from 1,949 commercial flocks, each consisting of thousands of broilers. For 1,315 flocks, only one sample of 23 chicks was collected from each flock. For 400 flocks, two samples of 23 chicks were collected. More than two samples were taken from each of the remaining flocks. The age of the tested samples ranged from zero (i.e., the day the chicks hatched) to 47 days, but the majority of them were between 28 and 34 days old.

We have illustrated the average NDV antibody titer of all the samples at each test age, from zero to 47 days, pooled into a single grand mean (fig. 10.1). The data suggest that chicks of all ages in each flock were protected from NDV (mean titer ≥ 3). However, when we examined

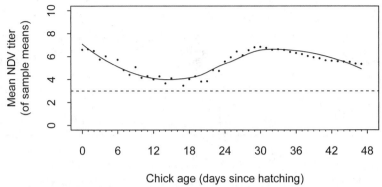

FIG. 10.1. Newcastle disease virus (NDV) antibody titer of chicks as a function of their age. Each dot represents the mean of all of the 23-chick samples for each sampling age. The solid curved line was fit to the data using local weighted regression and shows the average age-dependent levels of antibody in commercial broiler flocks.

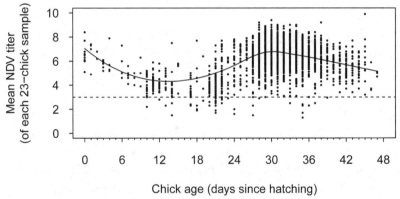

FIG. 10.2. Mean Newcastle disease virus (NDV) antibody titer of each 23-chick sample, plotted as a function of sampling age. Unlike in Fig. 10.1, each dot represents the mean of a single 23-chick sample. As in Fig. 10.1, the solid line was fit to the data using local weighted regression to show the average age-dependent levels of antibody in commercial broiler flocks.

each flock separately, we found 53 samples from 39 flocks with mean NDV titers < 3 (fig. 10.2).

However, due to high stocking density in chicken flocks, if even a few chicks are not protected, they can trigger infection in other chicks, even those that are producing relatively high titers of antibodies. In fact, the percentage of protected chicks within samples varied greatly among samples (flocks). In 578 flocks out of the 3,063 sampled that had a mean

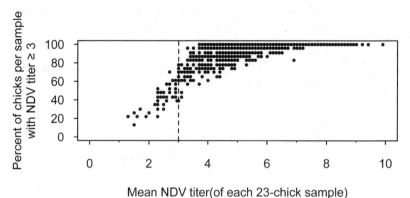

FIG. 10.3. Bivariate distribution of the hemagglutination-inhibition (HI) titer of Newcastle disease virus (NDV) antibody and the percent of NDV-protected chicks in each sample. The distribution shows the mean of each 23-chick sample, and the NDV-protected chicks are those with antibody titers ≥3.

antibody titer ≥3, fewer than 100 percent of their chicks were truly protected (fig. 10.3). These data suggested that it would be more important to estimate the percentage of protected chicks in each sample than the overall mean antibody titer to determine whether a flock is protected against NDV.

Observations from commercial flocks illustrated the importance of measurements other than the sample mean and the persistence of high antibody titers through time.[7] Although the average antibody titer of all flocks at each age was above the rate needed for NDV protection, in actuality many individuals within flocks were unprotected.

An experiment was done in which a combination of passive immunization with external antibodies and standard active (live) vaccination was given to overcome the temporal gap between the decrease in maternal antibodies and the increase in self-protective antibodies (see figs. 10.1 and 10.2), which would reduce the variability in protection among chicks in flocks. Broiler chicks were passively immunized beneath the skin (subcutaneously) with external antibodies when they were one day old; actively immunized with eye drops containing live vaccine when they were 10 days old; or given both immunizations.[8]

In chicks receiving both immunizations, mean titers of antibodies against NDV were >3 for the entire tested period (fig. 10.4, *panel A*). Chicks that received only the standard (active, eye drop) vaccine had a

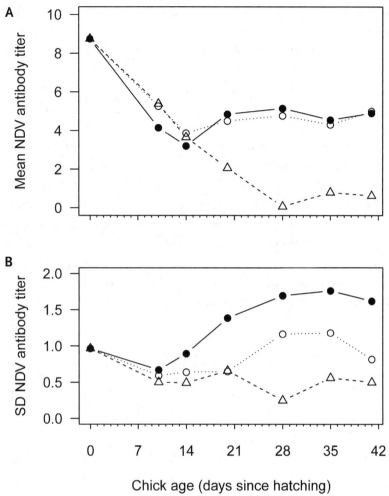

FIG. 10.4. Mean titers of antibodies against Newcastle disease virus (NDV) at different ages in chicks that received three immunization treatments (A) and standard deviation (SD) among replicate chicks within each of the treatment groups (B). The treatments received were passive immunization (subcutaneous) with external antibodies when one day old (△); active immunization (eye drops) with live vaccine when 10 days old (●); and both immunizations: passive on day 1 and active on day 10 (○).

similar response after 20 days and until 41 days, although antibody titers were somewhat lower on days 10 and 14. The mean antibody titers of chicks receiving only passive immunization on their first day were very similar to those of chicks that received combinations of passive and active immunizations, but only up until 14 days of age. Thereafter antibody titers in the chicks that were only passively immunized continued to decline (see fig. 10.4, *panel A*).

The combination of passive and active immunization not only led to higher antibody titers, but also led to more similar titers among repeatedly examined chicks. This is apparent in fig. 10.4, *panel B*, which shows that the SD of the mean antibody titer of these chicks was much lower than in the chicks that received only the active immunization.[9]

The combination of passive and active immunizations resulted in antibody titers ≥3 in all replicates (chicks) and smaller differences among them through 41 days of age, compared to the chicks receiving on the standard (active) vaccination (fig. 10.5). Thus, only the combination of the two immunizations provided all the chicks with protection against NDV for the entire 7-week rearing period (see fig. 10.5). This experiment demonstrates that the significance and reproducibility of differences in variation among individual replicates can be as important as differences among treatment means.

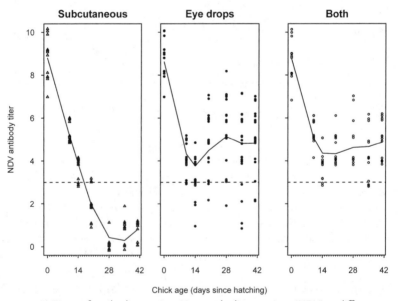

FIG. 10.5. Titers of antibodies against Newcastle disease virus (NDV) at different ages in individual chicks that received three immunization treatments: passive immunization (subcutaneous) with external antibodies when one day old (Δ); active immunization (eye drops) with live vaccine when 10 days old (●); and both immunizations: passive on day 1 and active on day 10 (○).

CASE STUDY 2: CANCER IMMUNOTHERAPY

Inbred mice (and other inbred "model organisms" in laboratory studies) from the same line are genetically identical. However, even when they are the same age, reared in the same room, given the same food, and treated by the same technician, mice from any given inbred line always differ to some extent in their responses to any treatment, resulting in a certain level of variation around the group's average.[10] In some cases, a substantial difference between treatments will not be found to be statistically significant because this within-group variance exceeds the variance between treatment groups.[11]

The high level of variance within groups (i.e., among individual replicate mice within treatment groups) occurs because the replicate mice are not actually identical within and among experimental groups. The best way to change the relative amount of within- and between-group variance is to identify and statistically account for additional variables (covariates) that are contributing to the higher within-group variance.[12] Many researchers think that increasing sample size also can reduce the estimate of the within-group variance, but for any "reasonable" sample size (e.g., ≥30), that is unlikely.[13]

Many experiments are done every day to test the efficacy of treatments for cancer. The local "microenvironment" of a cancerous tumor consists of cancer cells and various other cell types, including immune cells ("leukocytes"). These immune cells can acquire characteristics of tumor support when they come under the influence of secretions by the tumor cells themselves.

In this second case study, an experiment with micro-particles (MPs) carrying immune-cell inducing molecules to "re-direct" the intra-tumor leukocytes toward an anti-tumor response was repeated. The molecules loaded on the MPs induce changes in the secretion of the signals ("cytokines") by the leukocytes. A previous study had resulted in a statistically significant reduction in the rate of tumor progression in mice treated with MPs co-loaded with two inducing molecules. However, in a second experiment,[14] a similar effect was observed but one that was not statistically significant when the data were analyzed with a simple model that included only the effect of the MP treatments.[15] However, a significant effect of MP treatment was found when the effects of cage and four covariates were included in the model.

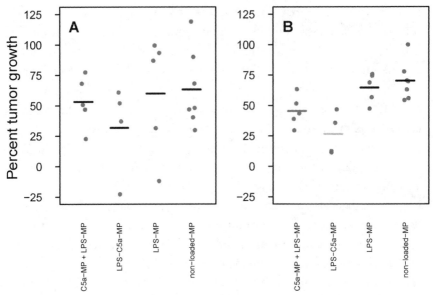

FIG. 10.6. Differences in variation in tumor growth among replicate mice subjected to four different treatments. *Panel A:* Scatterplot of individual replicates within each treatment; the sample mean in each treatment is indicated by a horizontal line. *Panel B:* Scatterplot of individual replicated adjusted for the effects of cage and four other covariates; the sample least-squared (LS) mean for each treatment is indicated by a horizontal line. The different shading of the horizontal lines indicates significant differences ($P < 0.05$). C5a = complement component 5a; MP = microparticle; LPS = lipopolysaccharide.

The simple mean of percent tumor growth of mice treated with MPs co-loaded with two different molecules (LPS and C5a-pep), although lower (31.8) than in the other treatments (53.2, 59.9 and 63.2; fig. 10.6, *panel A*), did not differ significantly from them. It was apparent that the relatively large differences among mice within treatments (see fig. 10.6, *panel A*) were reduced (fig. 10.6, *panel B*) by accounting for the variation in covariates, including rearing cage, pre-injection tumor development rate (days from seeding, tumor volume at injection), body weight at injection, and weight change after injection.[16] Concomitantly, the reduction in the error (residual) variance from 1,259.8 to 362.9 increased the ratio of among-group variance relative to the within-group variance (F ratio), leading to a statistically significant effect of MP treatment (i.e., P decreased from 0.5514 to 0.0441).

The analysis of the data using the model with covariates yielded "least-squares" (LS) means of the treatments: means that are adjusted for random differences in the values of the covariates for mice allocated to the different treatments. The LS mean of percent tumor growth for the mice treated with the co-loaded MPs (26.6) was significantly lower than that for those treated with LPS-loaded and non-loaded MPs (64.6 and 70.2). The LS mean of the mice treated with a mixture of LPS-loaded and C5a-loaded MPs (45.5) was intermediate (figure 10.6, *panel B*).

The extended model also was effective in the case of the cytokine TNF-α, but to a lesser extent (the *P* value decreased from 0.18 to 0.05) because three covariates had no significant effects on this cytokine, and two others were marginally significant (0.10 > *P* > 0.05).[17] Thus, extending a model with covariates is not a "magic wand."[18] Rather, it is useful only when treatment means differ considerably but not statistically significant due to covariate-dependent great variation among replicates. In this case, the extended model worked well for tumor growth and for the cytokine TNF-α but not at all for the chemokine CCL-2.

CONCLUDING REMARKS ON THE IMPORTANCE OF COVARIATES

The results of the case studies emphasize the importance of variation among individual replicates within treatment groups and of independent factors that may affect the responses to the treatments as covariates. Determining these covariates, recording their levels for each replicate, and including them in the statistical model used to analyze the data can result in two statistical benefits: lower residual (error) variance and unbiased estimates of treatment means (LS means). Reporting on such covariates, and whether they significantly affected the response variables, should be a *standard component* in any scientific publication. It will help those who run similar studies to improve their statistical analyses by identifying similar covariates and may also facilitate more correct meta-analysis studies.

NOTES

1. This is changing, however, with the increasing application of "personalized medicine": tailoring treatment to a particular person's genetic make-up.

Such personalized approaches remain quite expensive and as of this writing (2015) are available only to the very wealthy or to individuals enrolled in clinical trials.

2. Egea et al. (1991); Kruskall et al. (1992); Desombere et al. (2005).

3. Hamal (2006); Gharaibeh et al. (2008).

4. Berg and Meulemans (1991); Gharaibeh and Mahmoud (2013).

5. The antibody level is determined by a hemagglutination-inhibition (HI) test that results in titers ranging from zero (no antibodies that induce HI) to 12 (Maas et al. 2003; Vui et al. 2004).

6. Data were obtained from the Poultry Health Laboratories and Israeli Egg and Poultry Board.

7. Additional analysis showed that up to eight days of age, all the tested chicks were 100 percent protected because the chicks still had maternally derived antibodies. The half-life of these antibodies is 3–4 days, and their average level decreased from 6.5 on day zero to about 4.75 at 15–20 days of age (see figs. 10.1 and 10.2). Consequently, from day 10 onward, there were samples with less than 100 percent antibody protection, and the average proportion of such samples decreased to its lowest level at around 18 days of age. Following vaccination with attenuated NDV vaccine, a routine procedure in all chicken flocks in Israel, antibodies are produced by the chicks themselves. Consequently, both the average mean titer per sample and the percentage of protected chicks peak at around 30 days of age, with very slight reductions observed thereafter. However, at all ages >10 days, there were samples with <100 percent protected chicks. Over this range of ages, at least one non-protected chick (titer < 3) was found in 424 (≈22 percent) out of the 1,949 flocks tested. Of these 424 flocks, 405 had a mean antibody titer > 3, another demonstration of the limited value of titer mean as an indicator of the protection of any given flock against NDV.

8. Subcutaneous injection was with 18-HI units of anti-NDV immunoglobulin Y (IgY) at one day of age. IgY is the avian homologue to mammalian immunoglobulin G (IgG), which is the major antibody produced in response to infection or vaccination. IgY is transferred from the hen into the egg yolk, is absorbed by the embryo, and confers protection during the first days after hatching. Eye drops administered at 10 days of age included live attenuated NDV (Yosipovich et al. 2015).

9. These differences in homogeneity were examined by post-hoc tests (Bartlett's, Levene's, O'Brien's and Brown-Forsythe's) and found to be highly significant at 20 and 42 days and marginally significant at 14, 28, and 35 days. In the passive immunization treatment no self-production of antibodies was evoked, and the SD values were the lowest at 28, 35, and 41 days (fig. 10.4, *panel B*).

10. Leek et al. (2010).

11. That is, $P > 0.05$; see Chapter 9.

12. Such covariates are not germane to the objectives of the experiment— statisticians often refer to them as "nuisance variables"—and the goal of a

well-designed experiment is that the "signal" of the experimental treatment exceeds the random "noise" of the nuisance variables. Ecologists doing field experiments deal with such covariates all the time, but they often are unexpected in the much more controlled environment of a laboratory. Either way, once such factors are identified and recorded during the experiment, by including them in the statistical model, their effects are accounted for in the statistical model instead of inflating the "residual" (or unexplained) variance. See Gotelli and Ellison (2012: Chapter 10) for analytical treatment of such covariates in standard analysis of variance (ANOVA) models.

13. Increasing sample size, however, will reduce the standard error (SE) of the mean, which is calculated as $\dfrac{SD}{\sqrt{n}}$. A smaller SE will lead to improvement in significance tests (lower probability of a Type I error; see Chapter 9) and smaller confidence intervals (Ellison and Dennis 2010).

14. C57BL/6 inbred mice were divided into cages and tagged with numbers running from 1 to 8 within each cage. Melanoma cancer was induced in the right flank of each mouse by subcutaneous injection of autologous B16-F10 cells. Tumor growth and mice weight were monitored daily. Upon reaching the tumor volume that had been determined a priori for experimental treatment (100–200 mm^3), the mice were injected intra-tumorally with one of four treatments: MP loaded with lipopolysaccharide (LPS), a potent inducer of immune response; MP co-loaded with LPS and a peptide analogue of complement component 5a (C5a-pep); a 1:1 volume mix of MP loaded with LPS and MP loaded with C5a-pep; and non-loaded MP (control). Treatments were assigned randomly and divided equally between cages before the trial began (for example, mice numbers 1 and 2 of each cage were treated with non-loaded MP). Two days after treatments, mice weight and tumor volume were recorded and the tumors were harvested and tested for cytokine expression levels. We have presented the results of calculated tumor volume growth after treatments of tumor growth and a representative cytokine and chemokine (see fig. 10.6; the ANOVA tables are shown in the following two endnotes).

15. We normally illustrate the statistical analysis of a designed experiment, such as this one with MP treatments of mice, using an ANOVA table. Here we have illustrated (table 10.1) how the variance is partitioned among the different treatment groups (the "treatment" source) and within groups (the "error" source). Degrees of freedom (df) provide information about the number of treatment groups and the number of individuals. The remaining terms provide information about the deviation of the individual means from the overall grand mean (the sums of squares), the deviation relative to the sample size (the mean square = the sums of squares / df), and the ratio of the among-group variance to the within-group variance (F ratio = mean square among groups / mean square within groups). The *P* value gives the probability of observing a significant effect if there really is no effect of the treatment (i.e., the probability of committing a Type I statistical error; see Chapter 9).

It is calculated based on the expected value of the F ratio for the given degrees of freedom if the null hypothesis is true. See Gotelli and Ellison (2012: Chapter 10) for detailed explanations of a wide variety of ANOVA designs and how to calculate the various terms.

Table 10.1

Analysis of variance table illustrating the statistical analysis of data shown elsewhere (see fig. 10.6, *panel A*)

SOURCE	DF	SUM OF SQUARES	MEAN SQUARE	F-RATIO	P
MP treatment	3	2,736	912.3	0.72	0.55
Error	17	21,416	1,259.8		

Note: The model fit includes only the effects of the micro-particle (MP) treatment on the percentage of tumor growth.

16. Table 10.2.

Table 10.2

Analysis of variance table of the extended model (LS means plotted in fig. 10.6, *panel B*)

SOURCE	DF	SUM OF SQUARES	MEAN SQUARE	F RATIO	P
MP treatment	3	4,245	1,415.2	3.90	*0.04*
Cage	3	7,059	2,353.3	6.48	*0.01*
Days to injection	1	5,870	5,870.0	16.17	*0.002*
Tumor at injection	1	2,852	2,852.3	7.86	*0.02*
Mouse weight	1	1,783	1,783.5	4.91	*0.05*
Percent of weight change	1	3,326	3,326.0	9.16	*0.01*
Error	10	3,629	362.9		

Note: Includes the effects of cages and four mice- and tumor-related covariates, leading to a nearly eight-fold reduction in the error sums of squares. As a result, the F-ratios are much larger, and the P values are smaller. Statistically significant P values (i.e., $P \leq 0.05$) are set in *italics*. MP = micro-particle.

Additional statistical aspects of the extended model include fewer error degrees of freedom (hence less "sensitive" tests due to higher significance-threshold values for the test statistics) and the estimation of LS means for each treatment that may change the sums of squares and mean squares.

17. Table 10.3.

Table 10.3

Analysis of variance tables for effects of treatments on TNF-α and CCL-2, both with and without covariates

	SOURCE	DF	SUM OF SQUARES	MEAN SQUARE	F RATIO	P
1. TNF-α	MP treatment	3	0.04	0.01	1.80	0.18
	Error	17	0.12	0.01		
2. CCL-2	MP treatment	3	0.06	0.02	1.40	0.28
	Error	17	0.23	0.01		
1. TNF-α	MP treatment	3	0.06	0.02	3.85	0.05
	Cage	3	0.02	0.005	1.08	0.40
	Days to injection	1	0.02	0.02	3.73	0.08
	Tumor at injection	1	0.001	0.001	0.25	0.63
	Mouse weight	1	0.006	0.006	1.19	0.30
	Percent of weight change	1	0.02	0.02	3.34	0.10
	Error	10	0.05	0.005		
2. CCL-2	MP treatment	3	0.02	0.01	0.68	0.59
	Cage	3	0.05	0.02	1.53	0.27
	Days to injection	1	0.0001	0.0001	0.01	0.91
	Tumor at injection	1	0.01	0.01	1.03	0.33
	Mouse weight	1	0.01	0.01	1.32	0.28
	Percent of weight change	1	0.04	0.04	3.31	0.10
	Error	10	0.11	0.01		

Note: MP = micro-particle.

18. It is also important to note that including multiple covariates in a model can lead to overfitting and that repeated testing of different covariates may increase the probability of committing a Type I statistical error (see Chapter 9). Thus, selection and use of covariates requires considerable care.

Meta-analysis and the Need for Repeatability

## CHAPTER ELEVEN

Replication and Repetition in Systematic Reviews and Meta-analyses in Medicine

Leonard Leibovici and Mical Paul

Three questions about replication and repetition in systematic reviews and meta-analyses merit discussion: What is the degree of similarity between studies that supports their inclusion in a meta-analysis, and what does the degree of similarity tell us about the uses of the treatment effect measurement? When should replication of studies be stopped because a high level of confidence has been reached and resources should not be squandered? And should systematic reviews and meta-analyses (being secondary analyses of data) be repeated?

Systematic reviews (SRs) are reviews of the "best available," reliable studies focused on a specific research question. In particular, SRs apply clearly defined standards and criteria in order to select and synthesize studies in such a way that biases of the individual(s) conducting the reviews are minimized and the syntheses can be reproduced by other investigators.[1]

Systematic reviews are used frequently in the biomedical sciences to summarize answers to clear clinical questions that include definitions of the applied treatments (interventions), comparison treatments, relevant patients, and the measured clinical outcomes.[2] Most often, the studies included in SRs are randomized controlled trials (RCTs) that have repeated the same treatments in (usually) different situations. RCTs are preferred because randomization should result in an unbiased distribution between groups of known and unknown risk factors (other than the tested intervention) that might influence the outcome, leaving the intervention as the single difference between the groups. This design ensures that if a difference is shown in the outcome, it has been caused by the intervention. (See Chapters 9, 10, and 17 for additional discussion of experimental design).

A single RCT rarely is sufficient to provide reliable generalization to larger populations.[3] An SR of a large number of RCTs repeated on a variety of populations, together with a detailed understanding of the underlying cause-and-effect relationships (between, for example, the workings of a disease and the operation of a particular drug), can provide strong evidence to address key questions underlying an SR. Meta-analysis (MA) is a statistical method applied to the results gleaned from an SR that yields a single measure of the expected outcomes of repeated trials, along with an assertion of the confidence we have in that measure.[4]

RCTs are never identical: it is difficult to claim that one trial is an exact repetition of another, even if the protocol is identical. Patients differ from one place to another; interpretation and application of the protocol might be subtly different from one trial to another; the different times during which RCTs are conducted likely experience different environmental or clinical conditions.[5] Thus, we consider RCTs to be repeated trials but not exact replicates (see Chapter 1). In such repeated RCTs, conditions may differ, but the same treatments are applied and outcomes are measured.[6] An interesting question is this: what is the minimum number of RCTs needed to convince us that an effect exists? The minimum acceptable to most regulatory authorities is two. But in truth, the number should depend on the variability among patients, the treatment and underlying condition(s) tested, and other contextual variables: we agree with Cartwright that there is not a fixed "right" number.[7]

DEFINING EFFECTS IN SRS AND MAS

In SRs and MAs the effect can be defined very precisely, very broadly, or somewhere in between. However, the degree of specificity with which we define the effect size determines the meaning we can read into the effect. Two examples illustrate this challenge. First, cystitis is a superficial bacterial infection of the urinary bladder causing burning on urination without fever or other systemic manifestations. Some episodes might resolve without antibiotic treatment, and thus the question of whether cystitis should be treated with antibiotics is important. A SR and MA addressing this question yielded a positive answer: patients treated with antibiotics achieved a higher rate of cure than patients given a placebo or no treatment.[8] However, the results were not helpful to answer practical questions about which antibiotic drug should be given or how much should be given and for how long. To answer these questions we need first to answer more specific questions identifying the types of patients (e.g., men, women, children, and/or sub-groups thereof) and interventions: Is one antibiotic better than another? What is the optimum duration of treatment?[9]

Second, cancer patients with a low count of white blood cells are at increased risk of bacterial infections. We can ask whether prophylactic antibiotics lower the risk of infection and improve survival; but to yield practical answers we must focus on defined groups of patients (e.g. patients with acute leukemia) who received a specific drug.[10]

WHEN ARE THERE ENOUGH REPEATED RCTS?

A crucial question for experimental research is this: when has an experiment been repeated enough times that the results are convincing? More specifically, we can separate this question into two parts. First, is it theoretically possible to call for a stop to further original studies (and under which conditions)? Second, does this actually happen in practice?

A recent meta-analysis of 20 RCTs, which included 29,535 participants in total, examined whether vitamin D could prevent falls.[11] This study concluded with a high level of certainty that there was no effect of vitamin D on falling, and the confidence intervals for no effect ("futility") suggested that no further trials should be conducted for this intervention. An identical conclusion could have been reached a decade earlier. In 2005, after 8,878 patients had been studied in seven RCTs, the accumulated results had already crossed the futility boundaries. Nonetheless, more than 20,000 additional participants were studied in unnecessary RCTs, wasting valuable time and resources.

An SR and MA published in 2000 examined whether the addition of the anesthetic lidocaine (aka lignocaine) to an injection of the drug propofol could prevent pain; it could.[12] In the next 14 years, however, 136 new trials (including 19,778 patients) of the same treatment were conducted, with no new conclusions. More importantly, however, the conclusions of the 2000 SR and MA had surprisingly little impact on the design or reporting of the subsequent studies.[13]

In contrast, in 2004 we found abundant evidence that the addition of an aminoglycoside antibiotic to a beta-lactam antibiotic drug does not improve the outcomes of patients with sepsis.[14] The mean number of RCTs on this subject published between 1984 and 1993 was 3.2 per year, whereas it was 2.2 for years from 1994 to 2003. Reviewing this field again in 2014, we found that four more trials had been conducted on the subject—that is, 0.4 trials per year—in the intervening decade.[15] The earlier systematic review might have influenced the reduction in trials after 2013. Examining these three examples, we could expect the antibiotic SR to have had the least influence. It addressed a broad intervention as compared to the other two, in which the intervention was narrowly defined; besides, resistance to antibiotics has been increasing through time, which may render results from older trials irrelevant.

SHOULD SRS AND MAS THEMSELVES BE REPEATED?

Finally, we consider whether SRs and MAs themselves should be repeated (this is not the same as reproducing the analysis of a single SR/MA). Siontis et al. examined 73 MAs randomly selected out of all MAs that were published in 2010.[16] Two-thirds of the MAs overlapped with other, already published, MAs. The maximum number was 13 MAs performed and published on the same subject. Yet, SRs and MAs are secondary analyses, and there is no obvious reason to repeat them. In most SRs and MAs (and in all Cochrane reviews), reproducibility is imbedded in the design: two independent reviewers extract the same data and compare results.[17] Conflicting results are discussed with a third reviewer. Why would SRs and MAs be repeated or reproduced? There are good reasons and bad reasons.

Among the good reasons is the inclusion of new evidence. Every Cochrane review is updated periodically, but the new version replaces the old, and the update is performed by the original authors (unless they decline to update the old version). New trials demand updating of the reviews. Another good reason to repeat an SR and MA is to look for an outcome different from the outcome that was addressed in a former review. We stress that the main outcome should be one that matters to patients. If such an outcome was not used in an SR, repeating the SR based on a different (and more pertinent) outcome is justified. Similarly, an SR and MA should be repeated if the original analysis used flawed methodology.

In contrast, the need for another publication simply to pad a researcher's publication record, when there is no new evidence (or inconsequential new evidence), is a bad reason for repeating a review. We have seen a few examples in which SRs and MAs were repeated in order to contradict the results of another SR and MA. If an MA shows that a drug is at a disadvantage compared with other drugs (the relative risk for the bad outcome is greater than one, with narrow confidence intervals that do not include one), the easiest way to contradict the results is to reduce the sample size by including fewer articles. This can be done by including only articles published in English or including only articles that were reported according to very high standards. The denominator can be reduced, and thus the confidence interval increased, until the confidence interval contains one and thus the results are no longer significant. However, this selective filtering, while transparent and reproducible, may not be justifiable.

THE ULTIMATE VALUE OF SRS AND MAS

To sum up, RCTs are never similar enough to be considered identical replicates, but they are repeated studies, usually on different populations under different conditions. Comparable RCTs examine one or similar outcomes (based on a hypothesized cause-and-effect relationship), which is why comparable RCTs can be included in SRs and MAs. One would hope that when SRs and MAs show convincing results further repeated RCTs would be avoided, saving valuable resources. However, evidence to date suggests that this rarely occurs.

Although there are good and bad reasons for repeating SRs and MAs, reproducibility is imbedded in the creation of reliable SRs and MAs. Data are extracted by two independent researchers, and the resulting analyses are compared. The best reasons for repeating SRs and MAs are that new evidence has become available; a new outcome has become of interest; or there is a need to correct bad methods.

NOTES

1. See the Campbell Collaboration: What Helps? What Harms? Based on What Evidence? "What Is a Systematic Review?," http://www .campbellcollaboration.org/what_is_a_systematic_review/index.php, for a formal definition of systematic reviews and criteria for including or excluding studies from systematic reviews.

2. The consistent application of these criteria avoids many of the problems identified by Cartwright (2012; see especially her Chapter 4.1) with simple application and extension of the results of a single (or few) randomized controlled trials to a broader population.

3. Cartwright (2012).

4. Technically, an n percent confidence interval is the probability that the average *true* effect size (denoted by μ) lies between the lower (L) and upper (U) confidence bounds is equal to n percent. For a 95 percent confidence interval, that is $P(L \leq μ \leq U) = 0.95$. There are two practical difficulties with this. First, the results of each experiment (e.g., RCT) and of the meta-analysis provide *estimates* of μ (which we denote $\overline{X}$; see Chapter 9), not the true (unknown and unknowable) value of μ. Second, because μ is a true value, it is either in the interval [L,U] (with probability 1) or it is not (with probability zero). The 95 percent comes up because we do not know what μ is. So the correct interpretation of a 95 percent confidence interval is that 95 percent of the confidence intervals resulting from a hypothetical infinite number of repeated random samples (in which L and U themselves will vary randomly from sample to sample) will contain μ. See Ellison and Dennis (2010: Panel 1) for detailed discussion.

5. Unfortunately, data are sometimes manipulated to show a positive effect. This is another reason for the performance of more than one RCT, preferably by independent researchers.

6. This is similar to repeated sampling in monitoring studies (see Chapters 6–8).

7. Cartwright (2012: Chapter 2.3).

8. Falagas et al. (2009).

9. Milo et al. (2005); Zalmanovici Trestioreanu et al. (2010).

10. Gafter-Gvili et al. (2012).

11. Bolland et al. (2014).

12. Picard and Tramèr (2000).

13. Habre et al. (2014).

14. Paul et al. (2004).

15. Paul et al. (2014).

16. Siontis et al. (2013).

17. http://www.cochrane.org/cochrane-reviews.

## CHAPTER TWELVE

Clinical Heterogeneity in Multiple Trials of the Antimicrobial Treatment of Cholera

Mical Paul, Ya'ara Leibovici-Weissman, and Leonard Leibovici

A review of 24 randomized controlled trials on antibiotic treatment for cholera revealed that there was no true repetition among these clinical trials. Surprisingly, however, the results were consistent among trials. Meta-analysis is a valuable tool for combining heterogeneous trials and for addressing fundamental and conceptual clinical questions.

As we discussed in Chapter 11, meta-analysis (MA) is a statistical method applied to the results gleaned from a systematic review of the published literature that yields a single measure of the expected outcome of a set of trials along with an assertion of the confidence we have in that measure. Importantly, the studies included in a valid and informative MA are expected to be repeated studies, usually randomized controlled trials (RCTs), that were done on a variety of different groups of individuals (populations) using standard protocols and treatments and with repetition defined by similar patients, interventions, comparisons, and outcomes.

In this case study we examine whether this is indeed the case based on a Cochrane systematic review (SR) of antibiotics for treating cholera.[1] Despite demonstrating that none of the RCTs were identical, differing not only in patient characteristics (as might be expected) but also in antibiotics tested, setting, criteria for including or excluding individuals from the study, and other methods, we find that the MA of these heterogeneous trials was valid and informative. Our analysis sheds additional light on the importance of repeated experiments and cautions against applying strict meanings of replication in discussions of MA and RCTs.

WHAT IS CHOLERA?

Cholera is an acute diarrheal disease caused by the bacterium *Vibrio cholerae* (henceforth, cholera). Cholera has many serogroups, of which serogroups O1 and O139 cause the clinical disease.[2] The O1 serogroup is further divided into several biotypes, including the "classic" cholera biotype and the "El Tor" biotype.[3] The bacterium that causes cholera is transmitted to humans when they ingest contaminated water or food, and drinking water contaminated with cholera is a major mode of disease transmission in developing countries. Cholera is an epidemic disease, with potential for causing pandemics and establishing endemicity. The disease is currently endemic in Asia and some parts of Africa. Recently outbreaks have been reported in Zimbabwe, South Sudan, and Haiti and have spread from Haiti to the neighboring Dominican Republic.[4]

The clinical effects of cholera range from mild diarrhea to the life-threatening disease that causes rapid and severe dehydration. Diarrhea is caused by a toxin secreted by the pathogenic strains of cholera (the cholera toxin), which binds to the mucosal cells of the small intestine. The activity of the toxin blocks absorption of sodium and causes excretion of chloride and water, resulting in the typical cholerous watery diarrhea. Death may result from severe dehydration and electrolyte imbalance. However, with modern case management, including rapid diagnosis and institution of treatment, the mortality rate is relatively low. For example, in the recent outbreak in Haiti, of the 707,176 suspected cases, there were 8,614 cholera-related deaths (1.2 percent).[5]

The mainstay for treating cholera is rehydration. Fluids and electrolytes are provided either orally as oral rehydration salts solution or intravenously, and cholera is curable if rehydration begins early enough in the course of severe disease. In addition, antibiotics may help to cure cholera in two different ways. First, antibiotics can kill the cholera bacterium. Second, antibiotics could inhibit the production of toxins by the bacterium. Note that as of this writing, no drugs have been developed that target the toxin that is already bound to the mucosal cells in the small intestine.

In our Cochrane review (see endnote 1), we examined the additional potential benefit of using antibiotics to manage cholera. More precisely, we asked whether the killing of cholera and inhibition of protein (toxin) synthesis shortens the duration of diarrhea in cholera and reduces the total volume of diarrhea. Secondarily, we compared the effects of different antibiotics, as only some antibiotics (tetracyclines and macrolides) inhibit protein synthesis and therefore could inhibit toxin production. Here we discuss only the first objective: the potential benefits of using antibiotics, relative to placebo or no treatment at all, to manage cholera.[6]

SELECTING STUDIES TO INCLUDE IN THE SYSTEMATIC REVIEW AND META-ANALYSIS

In our SR and MA, we included RCTs that compared the outcomes of patients with microbiologically confirmed cholera treated with an antibiotic versus those of patients whose cholera was managed without antibiotics (those receiving a placebo or no-treatment controls). The trials could include both adults and children. The primary outcomes selected for the review were diarrhea duration and stool volume. These outcomes were selected based on consultation with a researcher and co-author of the Cochrane review who has vast experience in the treatment of cholera patients in Bangladesh, who considered these outcomes as those best correlated with the recovery and well-being of patients.[7]

We searched the literature broadly in an attempt to identify all published and unpublished RCTs on antibiotics used to manage cholera, without restrictions as to language or the year of the study. We extracted from each RCT descriptive data on its setting and methods; criteria for including or excluding participants; definitions of outcomes and methods to measure and assess them; characteristics of the study cohort; and all outcomes. We pre-defined several sub-groups for separate analyses. These included children versus adults; the cholera serogroup that was present in the study population; patients with severe disease versus those with milder cholera; and precise versus imprecise time points when outcomes were assessed.

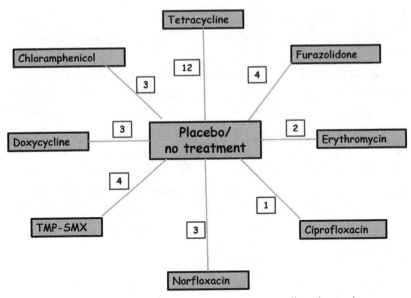

FIG. 12.1. Comparisons included in the systematic review. All randomized control trials (RCTs) included a placebo or no treatment arm and compared placebo or no treatment to different antibiotics (lines connecting large rectangles). The number of RCTs included for each comparison is shown in the small squares. TMP-SMX = trimethoprim-sulfamethoxazole.

RESULTS

A total of 24 RCTs were included in the comparison of antibiotics versus placebo or no treatment, comprising 32 comparisons, since some trials tested more than one antibiotic (fig. 12.1). Diarrhea duration was assessed by pooling the absolute mean difference between study arms.[8] Forest plots (figs. 12.2 and 12.3) were used to illustrate the results. In a forest plot, the estimated effect of the antibiotic in each individual study is depicted as a circle with 95 percent confidence intervals (CIs) extending as horizontal lines from the point estimate (see Chapter 11, note 4, for a primer on interpreting CIs). A vertical stack of the results from each of these individual studies creates the "forest." Finally, an overall estimate of the combined effect size from the meta-analysis is shown as a single circle with its 95 percent CI.

The meta-analysis suggested that treatment with antibiotics reduced the duration of diarrhea by 36.8 hours (with a 95 percent CI of 30 to 43.5 hours) relative to treatment with a placebo or no antibiotics at all. Inspection of the forest plot (see fig. 12.2) reveals that the results of all of the

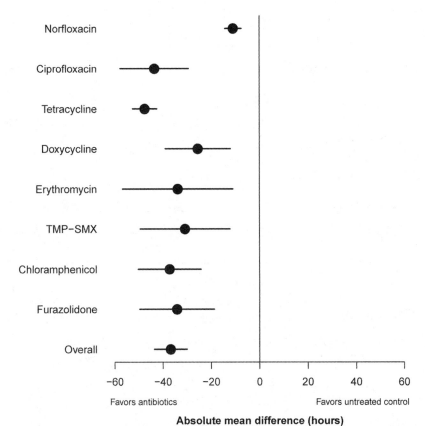

FIG. 12.2. Comparison between antibiotics versus placebo or no treatment, by duration of diarrhea (hours).

trials point in the same direction: a net benefit of antibiotics. Furthermore, statistical tests found no among-study heterogeneity for each type of antibiotics,[9] although there was some statistical heterogeneity among the different antibiotics.[10] The SR and MA also found similarly positive results of antibiotics on stool volume (fig. 12.3).[11] Overall, the vast majority of individual RCTs, whether measuring diarrhea duration or stool volume, had significant outcomes that paralleled the results of the MA.[12]

REPEATABILITY AND REPLICATION IN RANDOMIZED CONTROLLED TRIALS

The results (see fig. 12.2 and 12.3) suggest that many RCTs that apply identical (or very similar) protocols to diverse populations can yield

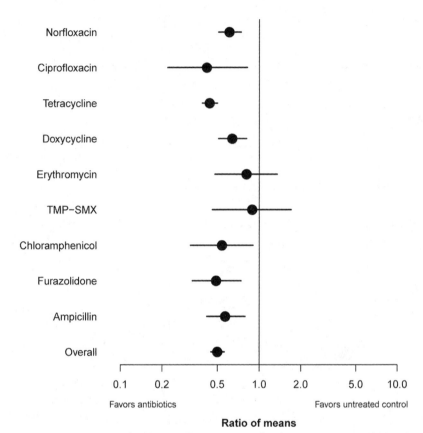

FIG. 12.3. Comparison between antibiotics versus placebo or no treatment, by stool volume (ratio).

similar outcomes and provide clear and uniform evidence for using particular treatments. But are the individual RCTs included in this SR and MA actually repeated RCTs?

Some of the characteristics of the RCTs are shown (table 12.1). Even with this small set of variables, it is difficult to identify two identical trials. The trials obviously examined different antibiotics, but they were also performed in different locations, included different patient populations because of trial-specific criteria for including or excluding patients, and used different strategies for hydration. In addition to the variables shown in the table, many other measurable variables were heterogeneous, including trial methodology, the methods for disease definition and outcome measurement, the actual severity of disease among included patients, cholera biotype, and resistance to the antibiotic used in the trial. Moreover, many other variables were unmeasurable or not reported, including the time from symptom onset to receipt of medical attention, the infrastructure of the health-care facility, the experience of caretakers, and infection control measures. Thus, although all of these RCTs test for the effect assessed in the SR, none can be identified as identical, repeated RCTs. The similarity among the results of the different trials initially may be surprising, given the large degree of heterogeneity among them.

One of the strengths of SRs and MAs, however, is the use of trials examining specific interventions to answer a principal question. Individual trials included in the meta-analysis tested the efficacy of a specific antibiotic, that is, they answered the question "Is this antibiotic effective in my setting?" The SR addressed the conceptual question of whether killing bacteria and stopping toxin production affects outcomes of patients with cholera. If we focus on a particular antibiotic, the different RCTs cannot be considered to be repeated RCTs. But if we think of each RCT as testing the effect of antibiotics, they can be considered to be repeated RCTs. In other words, by focusing on the principal question, the SR and MA place all of these similar RCTs on a level playing field. They allow the analyst to compare fruit to fruit as opposed to getting lost in the details of what kind of fruit was eaten (see Chapter 15).

Focusing on the big question does not mean that we can draw more specific conclusions about particular antibiotics, however. For example, because of the differences among trials, we cannot conclude that tetracycline is superior to norfloxacin.[13] This important distinction is often overlooked in critiques of evidence-based treatments derived from results of RCTs.[14]

WERE ALL THE RCTS ACTUALLY NEEDED?

Even though there is heterogeneity among the RCTs examining the same treatment(s) and outcome(s), it has been shown repeatedly that there is a limit to the information gained from repeating them (see Chapter 11). Can we determine, from this sample of 24 RCTs, if there was a point at which trials comparing an antibiotic to a placebo or no treatment became redundant and unjustified?[15]

In general, it is difficult to define a stopping point for repetition of RCTs because there are very few medical interventions whose effects are

Table 12.1

Selected characteristics of randomized controlled trials comparing the effects on cholera of an antibiotic versus placebo or no treatment

STUDY ID	ANTIBIOTIC[1]						COUNTY	SETTING[2]	FU[3]	PREVIOUS ABS[4]	SEVERE CHOLERA[5]	AGE GROUP[6]	FEMALES[7]	HYDRATION[8]	ILLNESS DURATION[9]	SERO-GROUP
	T	Q	TS	M	C	O										
Bhattacharya et al. 1990	0	1	0	0	0	0	India	En		X	I	C		ORS	<	O1
Burans et al. 1989	0	0	1	1	0	0	Somalia	En		X	I	B	X		<	O1
Butler et al. 1993	0	1	0	0	0	0	Peru	En	5	X	X	C	I		>	O1
Carpenter et al. 1964	1	0	0	0	0	0	India	Ep	7		I	C		Mix	>	O1
Dutta et al. 1996	1	1	0	0	0	0	India	En	5	X	I	C		ORS	<	O139
Francis et al. 1971	1	0	1	0	0	1	Nigeria	Ep	23	X	I	C	X		<	O1
Gharagozloo et al. 1970	1	0	1	0	1	0	Iran	Ep			X	B	X			O1
Hossain et al. 2002	1	0	0	0	0	0	Bangladesh	Ep		X	I	C		ORS	<	O139
Islam 1987	1	0	0	0	0	0	Bangladesh	En		X	I	C	I		<	O1
Kabir 1996	0	0	1	1	0	0	Bangladesh	En	5	X	I	A		ORS	>	O1
Karchmer et al. 1970	1	0	0	0	0	1	Pakistan	En	14		I	A	I	IV	<	O1
Khan et al. 1995	0	1	0	1	0	1	Bangladesh	En	4	X	I	C			<	O1
Lapeysonnie et al. 1971	0	0	0	0	0	1	Africa	Ep	8		I	B	I			O1

Study	T[1]	Q	TS	M	C	O	Country	Setting[2]	FU[3]	Previous abs[4]	Age[5]	Females[6]	Hydration[7]	Illness[8]
Lindenbaum et al. 1967a	1	0	0	0	1		Nigeria	Ep		I	C	I	IV	<
Lindenbaum et al. 1967b	1	0	0	0	1		Nigeria	Ep		I	A	I	IV	<
Lolekha et al. 1988	0	1	0	0	0	O1	Thailand	En	10	X	C	I		<
Pierce et al. 1968	1	0	0	0	0	O1	India	En		I	C	I	IV	<
Rabbani et al. 1989	1	0	0	0	0		Bangladesh	Ep	7	X	C	X	Mix	<
Rabbani et al. 1991	0	0	0	0	1		Bangladesh	Ep		X	A	X	Mix	<
Rahaman et al. 1976	1	0	0	0	0		Bangladesh	En		I	B	X		<
Roy et al. 1998	1	0	0	1	0	O1	Bangladesh	En	4	X	A	X	ORS	>
Usubutun et al. 1997	1	1	0	0	0	O1	Turkey	En		I	C	I		<
Wallace et al. 1968	1	0	0	0	1	O1	India	Ep	7	X	C	I	Mix	

1. Antibiotic tested (1) or not (0): Tetracycline (T), quinolone (Q), trimethoprim-sulfamethoxazole (TS), macrolide (M), chloramphenicol (C), other antibiotic (O).
2. Setting: Cholera endemic (En) or during an epidemic (Ep).
3. FU: Follow-up days in the trial (number of days).
4. Previous abs: Previous antibiotics excluded from (X) or included in (I) the trial.
5. Age group: Adults (A), children (C), or both adults and children (B).
6. Females: Excluded from (X) or included in (I) the trial.
7. Hydration: Oral rehydration salts (ORS), intravenous (IV), or a mixture of the two (Mix).
8. Illness duration: Less than 24 hours (<) or more than 24 hours (>).
9. Empty cells: Data not reported.

known beyond a doubt.[16] One also can debate whether the decision should be based on the results of individual trials or the results of an MA of all trials assessing the intervention. In some cases, an MA can show a significant effect of an intervention, even though the individual trials included in it individually did not find statistically significant effects of the intervention. In these instances MAs can draw attention to the futility of further trial repetition.

However, in our cholera case study, both the MA and all of the individual trials found that antibiotics were better than placebo. Why then, were so many RCTs performed using no treatment controls or a placebo? One reason could be that using a "convenience study design" is not harmful because the assessed outcomes are of minor significance relative to death from cholera.[17] Regardless, we conclude from our MA that future placebo-controlled trials of antibiotics for the treatment of cholera are not justified. Trials might be justified in the future if the pathogen changes its epidemic potential or pathogenicity or if resistance to all antibiotics increases to a degree that calls into question their efficacy.[18] New antibiotics should be compared to existing ones whose benefit has already been proven, not to a placebo.

A common criticism of MAs of the effects of antibiotics is that old trials are irrelevant because the specific antibiotics examined in these trials are no longer effective or used. We refute this for two reasons. First, MAs frequently examine a concept, as in our example. Old antibiotics were active at the time of the trial, as new antibiotics are active nowadays. To examine the effect of antibiotics, as such, on cholera it is correct to include old trials and thus to examine the overall evidence. Second, as resistance emerges with antibiotic use, it declines with discontinuation of its use. Such a phenomenon has been observed with tetracycline and chloramphenicol used to treat cholera. In the recent outbreak of cholera in Mexico, isolates were susceptible to tetracycline and chloramphenicol. Thus, it is important to include old trials in SRs for the future, when the antibiotics may become effective once again.

In summary, our case study of an SR and an MA of RCTs assessing the efficacy of antibiotics in the treatment of cholera has illustrated that even though RCTs cannot be repeated exactly, MAs that account for clinical and methodological heterogeneity are useful for addressing conceptual questions.

NOTES

We thank the Cochrane Infectious Diseases Group for their editorial assistance in preparing this Cochrane review.

1. In this SR, Leibovici-Weissman et al. (2014) included 23 randomized controlled trials that compared any antibiotics to a placebo or no treatment for adults or children with proven cholera.

2. A serogroup is a group of micro-organisms that are in the same species but differ with respect to the reactions they have to antigens (molecules on the surface of a cell that stimulate the production of antibodies).

3. Major differences exist between the pathogenicity and epidemic potential of the different biotypes.

4. Summaries of the current distribution, prevalence, and endemicity of cholera are in Sack et al. (2004), Barzilay et al. (2013), and updates provided by the World Health Organization for Mexico, Zimbabwe, and South Sudan (WHO 2009, 2013, 2014a).

5. WHO (2014b).

6. The comparison between different antibiotics found that azithromycin was significantly more efficacious than ciprofloxacin and erythromycin with regard to all outcomes (diarrhea duration, volume of diarrhea, duration of fecal excretion of pathogen, and amounts of rehydration fluids required). Tetracycline was compared to all antimicrobials except for azithromycin and was significantly more efficacious with regard to all outcomes. A comparison of different durations of antibiotic treatment revealed that longer than 24 hours' treatment resulted in shortening the duration of fecal excretion of cholera by about half a day.

7. Mortality, although perhaps ultimately the most important outcome, was defined in our SR and MA as a secondary outcome because it is so rare (especially in RCTs). Other secondary outcomes included the duration of excretion of cholera in stool; persistence of diarrhea beyond 48 hours; isolation of *Vibrio cholerae* from stool beyond 48 hours; and hydration volume. Note also that outcomes are directly analogous to indicators measured in monitoring studies (see Chapters 6–8).

8. The mean difference was calculated as the simple difference in mean values of the study arms, in our case study of the difference in the mean duration of diarrhea in days for antibiotics vs. untreated controls. This method is valid when all the trials measure a continuous variable with the same units.

9. Differences between trial results in a meta-analysis are quantified by assessment of heterogeneity or inconsistency. Two related terms are used in discussing heterogeneity among clinical trials: "statistical heterogeneity" and "clinical heterogeneity." "Statistical heterogeneity" refers to the statistical tests that measure clinical heterogeneity. The reasons for statistical heterogeneity are rooted in clinical or methodological differences between the trials. When heterogeneity is detected with statistical tests, clinical or

methodological variance between studies is indicated and needs to be investigated and explained. If heterogeneity is not detected statistically, clinical and methodological heterogeneity must still be investigated, as it might be masked by opposing direction or confounded effects. A discussion of such "interaction terms" can be found in Gotelli and Ellison (2012: Chapters 9 and 10).

10. Specifically, the greatest effect was observed in the pooled summary of RCTs comparing tetracycline versus placebo or no treatment (absolute mean difference of 47.4 hours (95 percent CI = 42.4–52.4) in favor of the antibiotic.

11. For stool volume, effect magnitude was calculated as the ratio of means between arms, because the trials reported the measurements of stool volume variable either as absolute volume or volume per weight. Overall, antibiotic treatment reduced stool volume by 50 percent (95 percent CI = 45–56 percent). Again, heterogeneity among RCTs was found among different antibiotics, but heterogeneity also was observed in some of the individual antibiotic comparisons (see fig. 12.3). A highly significant benefit was observed for all antibiotics except trimethoprim-sulfamethoxazole (TMP-SMX) and erythromycin. Tetracycline again had the greatest benefit (44 percent reduction in stool volume; 95 percent CI = 0.39–0.50), and the results for tetracycline were homogeneous among RCTs using this antibiotic.

12. There were no apparent deaths in any of the trials, although only six trials actually reported on deaths in the trials. All other secondary outcomes were significantly improved with antibiotics. The duration of cholera excretion in stools was shorter by 2.74 days with antibiotics (95 percent CI = 2.40–3.07 days). This implies shortening of the period of infectivity with antibiotic treatment, an important epidemiological consideration. Hydration requirements were reduced with an effect similar to the one observed for stool volume reduction.

13. Looking at the forest plots (see figs. 12.2 and 12.3), the pooled effect estimate for tetracycline appears to be larger than the effect estimate for norfloxacin, and the difference looks significant, since the 95 percent CIs of the pooled effects do not overlap. However, these effect estimates derive from different populations and different trial methods. Thus indirect comparisons (as in the example of tetracycline versus norfloxacin) are regarded as poor, essentially observational evidence, even though they are derived from RCTs. Further experiments, such as an RCT that tested tetracycline, norfloxacin, a placebo, and a no-treatment control on the same group (and then was repeated on different populations), would provide direct evidence for the efficacy (or lack thereof) of one antibiotic over another. See Chapters 9 and 10 for additional discussion of experimental design.

14. This also illustrates why the many repeated studies that are integrated into SRs and MAs provide much more evidence than the evidence provided by only one or a few studies (cf. Cartwright 2012).

15. As in Chapter 11, an RCT becomes redundant, unjustified, or even unethical if the benefit or harm of an intervention has been proven beyond doubt.

16. Statistical methods have been devised to deal with stopping individual trials before completion for harm or benefit. Strict criteria for follow-up assessment, significance testing, and sample size considerations must be fulfilled to justify early termination of a clinical trial. In fact, there are no formal guidelines for repeated trials testing an intervention. These decisions currently lie with the investigators and ethics committees. Given the great clinical heterogeneity of trials, as in our example, one might say that there is no limit to the repeatability of testing an intervention, since trials are inherently different. However, individuals whose lives depend on the development of new treatments likely are unwilling to wait forever to obtain the results from infinitely many RCTs.

17. A convenience study design is one in which subjects are chosen out of convenience (for example, they live close to the clinic) rather than being chosen at random.

18. Antibiotics lose effectiveness with time, because their use induces resistance. Resistance to tetracyclines, TMP-SMX, and chloramphenicol has already been detected in populations of *Vibrio cholerae* (Kitaoka et al. 2011).

*The Role of Metadata in Creating
Reproducible Research*

## CHAPTER THIRTEEN

Reliable Metadata and the Creation of Trustworthy, Reproducible, and Re-usable Data Sets

Kristin Vanderbilt and David Blankman

The volume and pervasiveness of data used for reproducing analyses and large-scale syntheses are increasing rapidly. All data that are re-used, modeled, or synthesized anew must be accompanied by metadata that establish their validity and trustworthiness.

Science has become a data-intensive enterprise. Data sets are commonly being stored in public data repositories and are thus available for others to use in new, often unexpected ways. Such re-use of data sets can take the form of reproducing the original analysis, analyzing the data in new ways, or combining multiple data sets into new data sets that are analyzed still further.

A scientist who re-uses a data set collected by another must be able to assess its trustworthiness. Trust in a data set refers to the degree to which a scientist is confident that (1) she understands the data set sufficiently to correctly re-use the data set in her own analysis and (2) the data accurately reflect the phenomena they purport to represent. The latter dimension of data trustworthiness, emphasized in this chapter, is a function of the quality assurance / quality control (QA/QC) process employed by the data creator and how well it is documented.[1]

Metadata are data about data. Metadata explain in language readable by machines or by people how to read and interpret the associated data set. In this overview, we review the types of errors that are found in metadata referring to data collected manually, data collected by instruments (sensors), and data recovered from specimens in museum collections. We also summarize methods used to screen these types of data for errors. We stress the importance of ensuring that metadata associated with a data set thoroughly document the error prevention, detection, and correction methods applied to the data set prior to publication.[2]

This chapter provides background material for the following one (Chapter 14), which discusses how to capture so-called provenance metadata for high-volume, high-frequency sensor data. Our focus in these two chapters is on ecological data (including natural history data) because these data are more complex and heterogeneous than data from many other scientific disciplines. The issues we identify for ecological data are, however, similar to issues found for data collected by molecular biologists, geneticists, physicists, and many other scientists.

SOME CHALLENGES FOR USING AND RE-USING ECOLOGICAL DATA

Many ecological and environmental observations are by their very nature not repeatable (Chapters 6–8). The exact circumstances (e.g., time of day, temperature, site history, and associated organisms; Chapter 16), and even the location (Chapters 4, 5, 15), of an observation is unique and will never exist again. The same variable, measured in different places by different people, may not be directly comparable to the one already observed due to methodological differences. Adding to the complexity of ecological data is their variability in spatial scale (e.g., from the microscopic to the landscape in extent) and temporal scale (e.g., from intervals of less than a second to more than a decade).

Ecological inferences and forecasts also transcend biological systems and take advantage of data from a range of academic disciplines and other sources. Answering questions about effects of climatic change, economic impacts of invasive species, and linkages between biological diversity and the services that humans derive from ecosystems uses data from many sources.[3] Ecologists need to discover, integrate, and analyze data from a wide range of topic- and discipline-specific data repositories that can be located anywhere in the world.[4]

When a scientist repeats an observation or experiment (Chapters 3–10, 15) or uses data from multiple studies in systematic reviews or meta-analyses (Chapters 3, 10, 11, 12), it is essential that data set documentation enable him to understand the complexities of the data and trust that the data are reliable and of high quality.[5] How do ecologists determine if the data they have acquired from a repository—data that were collected by another scientist—are indeed trustworthy?

ASSESSING THE QUALITY OF METADATA AND DATA

Metadata are the primary vehicle by which a secondary user can assess whether a data set is trustworthy and will meet her needs. At a minimum, for a data set to be suitable for a particular use case, the metadata accompanying the data must satisfy basic questions about the data: the who, what, when, where, and why of the data set.[6] Metadata usually are generated using a domain-specific metadata standard that specifies the content needed in the metadata file.[7] The completeness of a metadata file has an impact on the trustworthiness of a data set. Metadata lacking a description of data collection methods, for instance, may render a data set useless for some applications: A scientist conducting a meta-analysis of the effects of nitrogen enrichment on plant communities may not be able to use a data set about fertilizer effects on plant growth if the amount, type, and frequency of fertilizer application are not documented. As an example from museum collections data, if a researcher wanted to create a plant species list for Slovakia, herbarium specimens listing the collection location as "Slovakia" would be adequate.[8] However, the lack of more precise location information would make these data unfit for use by a scientist interested in studying the distribution of plant species within Slovakia.

The quality of the data records themselves, and thus their trustworthiness, can be compromised at many stages in the research process (table 13.1). Researchers must take steps to prevent the collection of "bad" data and ensure that the data that are collected do not contain errors. The procedures used for QA/QC of the data set (e.g., prevent, identify [flag], or correct errors) should all be detailed in the metadata.[9]

During the planning phase of a research project, field researchers can lessen the potential for errors by properly designing data sheets

Table 13.1

Quality considerations and activities to avoid collecting "bad" data or introducing errors or uncertainties into a data set

RESEARCH STAGE	QUALITY CONSIDERATION	ACTIONS TO AVOID ERRORS
1. Planning	Anticipate sources of errors, and manage project to avoid them	Design data sheet or handheld computer application to facilitate accurate and complete data collection
2. Data collection	Ensure consistency between data collectors	Train data collectors; document methods for reference in field
3. Data entry	Avoid transcription errors	Use data entry applications that screen for mistyped data
4. QA/QC	Identify and correct or flag questionable data	Implement sanity and range checks; use lookup tables
5. Data transformation/ analysis	Consider what a secondary data user will need to know to understand how the data were derived	Document assumptions and analytical steps

Note: Ensuring the trustworthiness of data requires vigilance throughout the research process, from planning the field campaign through data collection and analysis. Documentation of quality assurance / quality control (QA/QC) procedures throughout the research process is essential.

(or applications for data capture used with handheld computers) in advance of the field campaign. The data sheet should define what data are to be taken and how they are to be recorded on the data sheet.[10] Anticipating sources of error and designing the data collection instrument(s) to reduce them is preferable to post-data-collection "best-guess" interpretation by the individuals (often nonbiologists) who actually code and enter the data into spreadsheets or databases. Training data collection personnel (typically biologists) to ensure that everyone is taking measurements in the same way also reduces the likelihood that data will vary systematically between collectors.[11]

During the data digitization and QC phase of a research project, a variety of methods are used to detect and correct errors. Transcription errors such as mis-keying a number (e.g., 150 percent versus 15 percent) or transposing a letter in a species code (e.g., BORE4 vs. BOER4), can often be detected. Validation methods, such as using look-up tables to match species names as they are digitized to a master list of species from the study area, help identify errors at the point of entry, where they can be corrected most easily. Range checks, which also can be pro-grammed into a data-entry application, ensure that measurements fall within an acceptable interval. For example, a range check of valid values from +5 to +10 will prevent entering +55 rather than +5.5. Once the data have been entered, scripts (e.g., in SAS[12] or R[13]) can produce graphs that can help detect outliers. The data may also be summarized to, for ex-ample, identify missing or mis-labeled experimental units.[14] Carefully proof-reading entries and performing so-called sanity checks for "rea-sonable" outcomes or impossibilities can also help identify outliers or errors. For example, a Douglas fir tree that is observed to be dead in one year should not be censused the following year as a living Subalpine fir tree. Use of these and other standard "best practices" for preparing data to share will improve the quality and trustworthiness of a data set.[15]

Errors that are much more difficult to locate and correct are errors of omission.[16] For example, if a data collector fails to record the sex of a captured vole, it is impossible to obtain that datum once the animal has been released (unless it is tagged and subsequently recaptured). A broader characterization of errors of omission is the failure to record important contextual information that could help a subsequent user of the data understand anomalous (but not erroneous) records (see also Chapters 4, 5, 15). For example, rodents may become very abundant one year for no reason noted in the data set. Rodent population size is, how-ever, very responsive to the availability of food. If the year when rodent population size was large corresponded to a year when seeds were very abundant, the apparent anomalous size of the population would no longer be atypical.[17]

Raw data may go through several QC steps or be transformed in one or more ways as they are analyzed or prepared for publication.

Documentation of these steps is essential to make the meaning of the data transparent to a would-be secondary user.[18] Consider, for example, a desert field station where precipitation is automatically monitored every half hour at 10 sites. When a precipitation gauge malfunctions, data for the malfunctioning gauge are replaced with data from the closest functioning gauge. The assumption that the interpolated (or replaced or modeled) data are good estimates for the missing data needs to be justified (see Chapter 14), particularly in this arid environment where precipitation is patchy. These details, essential to understanding how the data were created and what they represent, must be included in the metadata. Otherwise, a secondary data user will not be able to understand the assumptions and potential sources of error inherent in the data set. Even a data set as straightforward as average daily meteorological values based on data that are collected hourly requires explanation. How many missing values in a day, for instance, are allowable before the daily average is considered invalid?

DATA QUALITY AND DATA TRUSTWORTHINESS

Even if individual data points are carefully checked and the metadata thoroughly document the methods and the QA/QC process, other characteristics of a data set can make it difficult to use or less trustworthy. Examples are abundant from the US Long Term Ecological Research (LTER) Network's Ecotrends project, during which one analyst acquired long-term data from 26 LTER sites and synthesized them to illustrate long-term trends in climate and meteorology, population dynamics, net primary productivity, and the chemistry of lakes and streams at and across these sites.[19]

The LTER analyst frequently was confronted with data sets whose formatting made using the data challenging. Issues encountered that lessened confidence in the data sets included (1) lack of headers for data tables, resulting in ambiguous column contents; (2) inconsistent data types in one column (e.g., characters mixed with numbers), which made the data difficult to use with some software packages; and (3) inconsistent ordering of variables in data sets that were subdivided into separate files based on season or year (e.g., data collected in Year 1 were ordered year, plot, quadrat, whereas data collected in Year 2 were ordered plot, quadrat, year). Data sets managed in spreadsheet software packages sometimes also contained column or row averages or totals in addition to the raw data. Data re-users accidentally may include these summary statistics in their own analyses, not realizing that both raw and summary data are mixed in the data file.[20] Finally, large gaps in the data were unexplained in the metadata and were not flagged or filled.[21]

Overall, the lack of documentation and inconsistent or poor formatting of the data sets lessened trust in the data. By contacting the principal investigator (PI) for each project, the analyst was able to resolve many, but not all, of these problems.[22] The need to contact the PI, however, underscores the inadequacy of the metadata. The PI often was needed to explain the data collection methods and the study design, because these details frequently were absent from the metadata. Also absent from all of these data sets were the step-wise descriptions of analyses and transformations that the data had undergone (its provenance; see Chapter 14) to arrive at the format and values in the archived data file.

Although trust in a data set is conveyed primarily through documentation, less tangible factors also influence a scientist's confidence in a data set. The perceived expertise of the individual who collected the raw data and who produced and posted the data set also may be important to a secondary data user.[23] Scientists view data sets through the lens of their own laboratory or field experience. Ecologists reported that being able to visualize how data were collected and having collected similar data themselves helped them to assess the limitations of a data set, sources of potential error, and how well the data set reflected the phenomenon being studied.[24]

NEW CHALLENGES POSED BY BIG DATA

"Big Data" refers both to massive amounts of data and to their complexity.[25] One type of Big Data that has led to new challenges for creating accurate metadata and engendering trustworthiness is data streaming from environmental sensor networks. Such sensor networks are being used to sample intensively multiple sites such as lakes, remote forest plots, and the ocean floor at much greater frequency than would have been possible previously without vast numbers of field assistants.[26] Many different types of data can be collected simultaneously, from soil temperature to photosynthetically active radiation and carbon flux. When transmitted across the Internet, such data provide researchers with near-real-time information about an enormous range of environmental processes, all of which can be linked together by the time stamps in the data.[27]

Relative to the manually collected ecological data sets described earlier, these streaming high-volume data sets require additional quality checks and additional metadata to convey trustworthiness. Novel sources of error in data sets collected by automated sensors include sensor drift; damage to sensors; disruption of sensors due to lack of maintenance or repair; loss of power when batteries fail; and faulty data transmission.[28] Ecologists have reported that their trust in sensor data is a function of confidence in the type of sensor used; knowledge that sensors are regularly calibrated and tested in the field; and documented reliability of sensors.[29] This contextual information needs to be recorded and accessible to all data users, both present and future, and tied to explicit time points (time stamps) within each data set. Metadata tied to a time stamp can be referenced easily and used to learn if an unexpected pattern in a plot of the data is due to a known issue with a sensor. Individual data points in streaming data often are identified ("flagged") to indicate if the data are "good" or if they are in some way questionable.

Automated quality checks often are used to identify problematic data collected by environmental sensors. In addition to range checks and sanity checks, algorithms may compare data streams from different sensors to ensure that they relate to each other as expected.[30] Further, streaming data are likely to be processed with many steps, including calibration, quality checks, and gap-filling before the final data set is released and contributed to a data repository. One way to convey the processing steps to data harvesters and re-users may be through a provenance graph (see also Chapter 14).

UNIQUE ASPECTS OF DATA IN NATURAL HISTORY COLLECTIONS

Natural history collections of plant and animal specimens provide valuable records of their historical distributions (see also Chapters 4, 5). Each specimen (e.g., a pressed plant, animal skin, or pinned arthropod) typically is labeled with its scientific name, a detailed description of where it was collected (including latitude and longitude), the name of the collector, and the date of collection. Many natural history museums now are digitizing these valuable biodiversity data, and transcription of sometimes faded or illegible hand-written specimen labels or associated field notes may be challenging. Scientific names are easy to corrupt during transcription, although such misspellings can be identified and corrected by using standard look-up tables of names of genera and species in data-entry programs.[31] Ironically, the rapid digitization of museum collections can reduce the fitness for use of specimen-label data if only part of the information is digitized in the haste to post the data from the collections online.[32]

Georeferencing, the process of associating individual specimens with their collection locations, is of paramount importance to researchers studying biodiversity, biogeography, epidemiology, and responses of species to land-use and climatic changes.[33] Mapping species occurrence data can help identify errors in georeferencing. For example, specimens of North American terrestrial mammals that are mapped to coordinates in the Pacific Ocean should have their digital and specimen label records examined for errors. Sometimes, however, the geographic coordinates of a particular specimen may be accurate, even though they are well outside of its range, and detailed metadata can identify accurate, but otherwise suspect, occurrence records.[34] Trust in the accuracy of the positional information influences how a species occurrence record can be used (see Chapter 15).[35]

For some legacy specimens collected before either reliable maps or global positioning system (GPS) units were available, location information can be quite general (e.g., "New York" or "Glacier National Park"). Lacking additional information, the latitude/longitude point locations assigned to such specimens might represent mid-town Manhattan or the center of Glacier National Park. An assertion of precision, such as the positional accuracy as measured by a polygon defining city or park boundaries or the radius of a circle around the inferred georeferenced location, should be indicated in the metadata. Inclusion of these details in the metadata will help subsequent users determine if they can trust the specimen data for use in specific applications. Specimens that have been collected more recently usually have more precise locations measured by GPS units, yet these data also have degrees of accuracy and precision that should be noted (on the difference between accuracy and precision see note 20 in Chapter 9).[36]

Specimen identification is another factor that can affect the trust-worthiness of specimen data. Misidentification of organisms by the original collector is a perennial problem for natural history collections. Collections managers or curators can include in the metadata their confidence in the correctness of the identity of a specimen based on their knowledge of the expertise of the person who identified the specimen. Species names also change as the taxonomy and systematics of organisms are revised. These changes in names, and who was responsible for them, also should be part of the metadata for museum specimens. A specimen with detailed metadata about its original collection, identification, any subsequent changes in nomenclature, and even comments by a curator regarding the trustworthiness of the specimen, will represent a quality record for research use.

DISCUSSION

In his play *Julius Caesar,* Shakespeare wrote: "The evil that men do lives after them; the good is oft interred with their bones." The same, unfortunately, can be said of ecological and natural history collections data. Researchers rarely are rewarded for documenting their data, often relegating it to the status of an unwelcome chore. Even when they do provide detailed metadata, the metadata are usually given in the context of the original data creator's research, which may not be sufficient for someone trying to re-use the data for a different purpose.[37] Some journal publishers are now requiring data to be archived as part of the publication process, so generating metadata will become more commonplace as this practice spreads.[38] Whether these metadata are sufficiently detailed to satisfy secondary data users remains to be seen. Perhaps metadata also should be subject to coordination or peer review by scientists in the same field who will know what pitfalls and methodological details to look for and can determine if they are appropriately addressed in the metadata.[39]

The trustworthiness of a data set is largely communicated through detailed documentation, and metadata creators should be encouraged to take advantage of some little-used metadata elements, or alternatives to textual metadata, to capture all the needed quality information. Many metadata standards include fields in which the QA/QC process is to be described, but in practice these fields rarely are filled out. Some metadata standards contain fields for logging questionable data discovered by secondary data users, but these fields are also rarely used because there is no straightforward way for data re-users to register their data use experience. Mechanisms should be developed to incorporate feedback from individuals who have re-used the data into the metadata or a companion Web site.[40] Finally, mechanisms for capturing the many steps taken to analyze, transform, and perform QA/QC on a data set should be developed. A provenance graph (Chapter 14) that allows secondary users to reproduce the analytical workflow of the data creator may greatly increase the trustworthiness of a data set.

Our examples from data collected by hand, by sensor networks, and from natural history collections have highlighted the broad challenges and identified paths for creating high-quality, trustworthy data. The following case study raises some additional challenges and potential solutions for documenting the analytical processes of creating the data sets themselves.

NOTES

1. The QA/QC process refers to activities employed to prevent and identify errors in a data set.
2. See NISO (2004) for an overview of the concept of metadata, Michener et al. (1997) for a detailed description of metadata for ecological data, Doctorow (2001) for a cautionary critique of the concept, and White et al. (2013) for some approaches to improving metadata.
3. Examples include Costanza et al. (1997) on the value of ecosystem services, Pimentel et al. (2005) on the economic costs of invasive species, and the regular reports of the Intergovernmental Panel on Climate Change (http://www .ipcc.ch).
4. Michener and Jones (2012).
5. We reiterate that this is not an issue only for ecologists. For example, despite the relatively simple structure of genetic data, the probability and extent of sequence errors in GenBank has been demonstrated repeatedly (e.g., Krawetz 1989; Wesche et al. 2004; Hubisz et al. 2011).
6. Michener et al. (1997).
7. Metadata standards increase the consistency of metadata across data sets to support long-term data discovery, use, and improved digital curation. The metadata standards used in ecological and museum research take the form of a structured set of elements that describe the data set or specimen. For example, the Ecological Metadata Language (EML: https://knb .ecoinformatics.org/#tools/eml) includes fields that: (1) facilitate data-set discovery (e.g., title, abstract, keywords); (2) describe the research project (e.g., temporal and spatial coverage, methods, experimental design); (3) indicate

data-set accessibility (e.g., contact information, URL to data, data use restrictions); (4) describe the data structure (e.g., attribute names, units of measure, file format); and (5) provide supplementary information that helps
users assess data quality (e.g., QA/QC procedures, code used in deriving and
processing data, and notes from secondary users of the data regarding
anomalous data points they detected). See Fegraus et al. (2005) for more
about EML.

8. Herbarium specimens typically consist of plants pressed and mounted on
acid-free paper (11×17 inches, or approximately 28×43 cm). The herbarium
sheet label includes species name, collector name, collection location, and
date of collection.

9. Locating, identifying (flagging), and correcting errors in a data set each can
be done in a number of different ways. The most important commonality
among these processes is that the original data set should not be altered
when errors are found. Rather, errant data should be identified (labeled or
flagged) as suspect or an error in the data set, with the corrected value added
in a new column. The method for determining correct values (models, intu-
ition, substitution, etc.) should also be noted in the metadata. Gotelli and
Ellison (2012: Chapter 8) discuss data QA/QC in great detail.

10. Technicians sent into the field with notebooks and a list of variables to record will each create their own style of entering the data on their notebook's
blank pages. This inconsistency makes the data hard to enter, and handwriting a new data sheet for each experimental unit is inefficient and prone to
errors. A good data sheet will contain blanks for all data to be collected in
the field. For example, a plant phenology study seeks to quantify when
plants green up, form buds, flower, fruit, and senesce. Technicians are to
record the phenological stage of each individual plant encountered on a transect. The data sheet should contain the study title and blanks for technician
name, date, comments, species code, and phenological stage. The data sheet
also should contain the list of codes and definitions for phenological stages.
Without this information readily at hand, a technician might create additional hybrid stages or his or her own set of codes, leading to confusion at
the time of data entry.

11. Suppose that technicians are asked to measure the height of grass plants.
One technician may interpret this as meaning to the top of the clump of
grass leaves, while another may think that she is supposed to measure to the
top of the flowering grass stalks. Having technicians do some test measurements together before taking real data in the field helps eliminate differences in interpretation of methods.

12. SAS 9.3 Software.

13. R Core Team (2015).

14. Suppose that a grassland plant production study takes place at three sites,
and at each site there are three rodent trapping webs. Around each web are
four plots, each divided into four quadrats. The quadrat is the unit mea-

sured. Scripts are employed to ensure that the correct number of units exist for each site/web/plot/quad combination, because mislabeling a location is a common error in grassland, where all plots tend to look alike.

15. Hook et al. (2010).

16. Brunt (2000).

17. While it may seem unlikely that metadata about rodent population data would not include the reason for a sudden population explosion, metadata often do not include such details. It may not occur to a researcher to include this contextual information. Or she may provide only the sparsest of metadata because there is no mechanism for rewarding researchers for investing time in creating high-quality metadata.

18. For additional detail on the challenges attendant to this documentation, see Chapter 14.

19. Laney and Peters (2006); Peters et al. (2013). Ecotrends data are available online at http://www.ecotrends.info.

20. Note, however, that such row or column totals, averages, or ranges are often computed and examined during the QA/QC process. But including summary or synthesized data with the raw data can be confusing to a secondary data user.

21. Personal communication from the analyst who synthesized long-term data from 26 LTER sites.

22. The existence of close communication and coordination among LTER site PIs enabled the analyst to contact the PIs and obtain the additional information that was needed. However, in a great many situations, data for metaanalyses and syntheses are simply harvested from Web sites and there is little communication (or opportunity for communication) with the individual who collected the data and produced the data set.

23. Zimmerman (2008) reported that ecologists viewed data collected by a graduate student or professor as more trustworthy than data collected by a technician. Further, the US LTER network of research sites has had a data management mandate since 1980, and the favorable opinion of scientists, both within and outside of the LTER network, of LTER data is partially based on the LTER network's reputation for careful attention to metadata and data quality (e-mail from Diane McKnight, PI of Niwot Ridge LTER, Colorado, USA, regarding a session she led about data archives at the American Geophysical Union's Fall 2013 meeting).

24. Zimmerman (2008).

25. For a recent review of the concept of Big Data, see Diebold (2012). For a detailed discussion of Big Data in ecology and environmental science, see Hampton et al. (2013); Ellison et al. (under contract).

26. Porter et al. (2005).

27. The collection of such data and making it rapidly available to the public are central goals of the National Ecological Observatory Network (NEON: http://www.neonscience.org).

28. Campbell et al. (2013) discuss the issue of the quality and trustworthiness of Big Data being produced by NEON and the novel sources of error in data sets collected by automated sensors.

29. Wallis et al. (2007); results based on interviews with scientists at the Center for Embedded Networked Sensing (CENS: http://www.cens.ucla.edu).

30. Other examples of such methods include assessing the data for persistently repeated values, which may signal a bad sensor or network failure; internal consistency, such that data for related parameters make sense (e.g., total nitrogen measured by one sensor should not be greater than nitrate nitrogen, a component of total nitrogen, measured by a co-located second sensor); and spatial consistency to ensure that sensors near each other respond realistically relative to each other (e.g., two plots receiving water treatments at the same time exhibit temporally aligned increases in soil moisture); see Campbell et al. (2013).

31. Chapman (2005).

32. Digitization and data entry of most specimens is done with student or volunteer labor, and the emphasis has been on making specimen data electronically available, no matter how skeletal the record (personal communication from a senior collection manager at the Museum of Southwestern Biology, University of New Mexico). Ayelet Shavit (personal communication) has noted that the field notes of the person who collected the specimen are rarely interoperable with the digital records of a specimen, thus relegating to unindexed and often inaccessible file drawers important information about the habitat or other circumstances of collection, such as slope, local weather, associated species, and soil types.

33. Costello and Wieczorek (2014).

34. By convention, longitudes west of the "prime meridian" at Greenwich, England, and east of 180° (the approximate location of the International Date Line) are assigned negative numbers. A missing negative sign on a longitude can place a shrub endemic to the Chihuahuan Desert in North America in a forest in Asia.

Data from individual herbaria and natural history museum collections increasingly are being deposited into larger, aggregated databases such as that of the Global Biodiversity Information Facility (GBIF: http://gbif.org; Edwards 2004) to allow users from around the globe to access vast numbers of species occurrence records all at once. GBIF now indexes over 500,000,000 records (as of 15 November 2014). The data that GBIF harvests from a data provider are known to contain errors: in a 2011 survey of 35 GBIF data contributors, 57 percent responded that they could not guarantee the quality of the data they contributed and 43 percent declined to respond to the data quality question (Costello et al. 2013). To improve data quality, GBIF national nodes offer tools for automated screening of harvested data for errors (e.g., "species name not in national or international checklists," "minimum and maximum altitude reversed," "coordinates are out of range for species"

[Belbin et al. 2013]). Some errors, however, require input from a domain expert to correct. Having data exposed in a systematic format through GBIF makes the data visible and allows qualified scientists to assess and correct them.

Other records that may appear to be inaccurately georeferenced, however, are actually correct. For example, GBIF data on the locations of the tropical coastal tree *Rhizophora mangle* reveal a record in Missouri, which is far from the tropics and well inland. In fact, there is an individual of this tree species growing in the Missouri Botanical Garden, so the GBIF record has accurate coordinates. There is nothing in the metadata, however, to indicate that the fitness for biogeographic use of such a specimen is very limited. In the future, GBIF will have tools to improve feedback to the source data provider regarding errors and anomalies such as this one so that the original and GBIF records will be congruent. A platform to support continuous community evaluation and feedback on data quality will also be available (personal communication from a senior program officer, Inventory, Discovery, Access, GBIF Secretariat).

35. The Darwin Core metadata standard (http://rs.tdwg.org/dwc), used widely by natural history collections, includes fields for documenting the degree of trust the data provider has in the location coordinates and taxonomic identification of a specimen. These fields, if filled out, convey important trustworthiness information to a would-be data user.

36. Other quality issues related to georeferencing include false precision of a location due to the entry of an unrealistic number of significant digits in decimal degrees (e.g., a georeferenced longitude of 167.988888345 for a location described as "1 km N of Deep Well" on the specimen label); incorrect designation of the geodetic datum (the coordinate system used by the GPS), and incorrect interpretation of a textual description of a locality (Chapman and Wieczorek 2006). Ideally, metadata from the original data provider would indicate that the many sources of georeferencing error have been checked.

37. In fact, researchers themselves may want to re-use their own data and, lacking even the most basic metadata, may find such re-use all but impossible. We all have experienced staring blankly at our computer screens looking at long lists of data files named "snails," "snails2," "snails3," etc. , trying to find the one that we need. And if we do find the one we want (perhaps "good-snails"), the variable names ("X1," "X2," "X3," . . .) turn out to be uninformative. If reconstructing these data sets can be frustrating and inefficient, or even impossible, for those of us who created them, think how difficult it will be for someone who finds them on our Web sites and want to incorporate them into the synthetic model she plans to publish in *Science*.

38. Whitlock (2011).

39. For instance, publication of a biodiversity data set with complete metadata could be similar to publishing a journal article (Costello et al. 2013). Data could be submitted to a data journal, and the journal would perform

technical, editorial, and peer review on the data and metadata to ensure that they are of high quality before they are released to the public. Peer review would include ensuring that the metadata document the sources of uncertainty in the data as well as any relevant contextual circumstances that may influence how the data are interpreted. This mechanism would make the data set citable and ensure that it is trustworthy. Unfortunately, the number of data sets far exceeds the number of peer reviewers, and expecting review of any but the most significant data sets may be unrealistic. With the reality of limited resources for information management in ecological and museum communities, the responsibility and commitment to coordination will likely always lie with the researchers generating the metadata to consider how to communicate to a secondary data user the fitness for use of their data (Gerson 2007, 2013).

40. Wikipedia (http://wikipedia.org) and other wikis provide obvious models.

## CHAPTER FOURTEEN

Replication of Data Analyses

PROVENANCE IN R

Emery R. Boose and Barbara S. Lerner

The ability to understand and reproduce data analyses is enhanced by metadata that document exactly how the data are created and transformed. The value of such data provenance is demonstrated using the R statistical language.

Virtually all scientific data today are managed and processed (and, increasingly, created) using digital technologies. Typically the data take the form of data proper (individual pieces of information) and associated structured metadata (literally, "data about data"; see Chapter 13) that explain how to read (or machine-read) and interpret the data. The information contained in the metadata is typically focused on the larger context of a data set (who, when, where, and why) and the details of individual data files (what—e.g., file structure, variable names, data types) rather than how the data were created and analyzed.[1]

In contrast, the metadata that describe how scientific data are created and analyzed typically are limited to a general description of data sources, software used, and statistical tests applied and are presented in narrative form in the methods section of a scientific paper or a data set description. Recognizing that such narratives are usually inadequate to support reproduction of the analysis of the original work, a growing number of journals now require that authors also publish their data;

some journals also require publication of the computer software (scripts) or workflows (if any) that were used to produce the published results.[2] However, finer-scale metadata that describe exactly how individual items of data were created and transformed and the processes by which this was done are rarely provided, even though such metadata have great potential to improve data set reliability.[3]

In actual practice, scientists may use a wide variety of types of software (not all of which are scripting languages) and often combine a series of scripts or other tools to complete an analysis. But our focus here is not on the particular software used but rather on the detailed process metadata, called "data provenance," required to ensure reproducibility of analyses and reliable re-use of the data.

We use the term "data provenance" to mean the information required to document the history of an item of data, including how it was created and how it was transformed. The term "provenance" (from the French *provenir*, "to come from") is perhaps best known today in reference to the world of art, where it means "the history of the ownership of a work of art or an antique, used as a guide to authenticity or quality; a documented record of this."[4] By documenting an unbroken sequence of ownership and location, provenance can help to establish that a work of art is in fact genuine. This concept of an unbroken chain of evidence that leads back to the point of origin turns out to be very useful in the world of data. Data provenance also can be examined rigorously to verify where an item of data came from and how it came to have its current value.[5]

THE NEED FOR DATA PROVENANCE FOR REPRODUCIBLE AND REPEATABLE RESEARCH

There are a number of reasons why the original data and script alone may be insufficient to guarantee that a data analysis is reproducible. One reason has to do with the inherent malleability of software. The ability to easily change the behavior of software to do something a bit different is what gives it great power and flexibility. This flexibility is enhanced in software environments such as R, Matlab, and Python, which, as interpreted software languages, are specifically designed to encourage interactive data exploration.[6] Unlike compiled software languages such as Fortran, C, and Java, which use a classical edit-compile-execute cycle, interpreted languages and their associated environments allow researchers to use a combination of pre-defined packages, user-created scripts, and commands selected from pull-down menus or executed individually or in small chunks directly from a console interface. Thus, it can be quite difficult to determine exactly what the researcher did and which version of her script was used to produce a particular result. Whereas most commercial software and software downloaded from the Internet has a clear version number, the same is rarely true for ad hoc scripts used for scientific data analysis. The exact version of the script used, including commands entered at the console or selected from pull-down menus, is an essential element of data provenance.

Another reason that data provenance is needed is that the outcome of running a script may not be deterministic. Rather, its execution can be influenced by the computing environment, including the version of hardware, operating system, and application software used; these pieces of information also are essential elements of data provenance. Execution also may be influenced by inputs that are not pre-determined in the script itself, including inputs from users, external sources (such as Web services), and stochastic processes (such as random numbers used for statistical sampling), or by unintentional errors, such as the incorrect interleaving of concurrent operations.[7] More generally, a script (or the source code in a particular computer language) helps us to understand what the software might do; it does not tell us what the software actually did during a particular execution.[8] For the latter—another crucial component of data provenance—we need the information contained in an execution trace: a record of all inputs, intermediate values, and final values, as well as error messages and warnings returned from the runtime environment.

Reliable and trustworthy scientific data analyses should be reproducible: applying the same analysis to the same input data should yield identical results (see Chapter 1). In reality, however, slight differences in results may occur. For example, representation errors may arise because different types of hardware and software store and use different numbers of significant digits. At a more profound level, because of the multithreaded nature of modern operating systems, it is not possible to exactly replicate the state of the original computer, even if identical hardware and software are used. If important discrepancies occur when attempting to reproduce computed results, data provenance can help researchers identify their causes.

One might also argue that data analyses should be repeatable (see Chapter 1). By analogy with monitoring (see Chapters 6–8), applying the same analysis to different but comparable input data should yield comparable results. But the extent to which an analysis is repeatable depends on the type of analysis. Just as the sensitivity of a model to different inputs depends on the structure of the model and its parameterization, so obtaining identical or radically different results from comparable but different inputs could provide reasons to investigate the underlying scripts. Knowledge of the programmer's intentions and access to the source code may help us to understand the results, but software is notoriously difficult to write well, and analyzing software for correctness is a notoriously difficult problem. In such cases the detailed execution trace provided by data provenance can improve our understanding of what the script actually did and can help us to detect errors.

CHALLENGES IN THE CREATION OF DATA PROVENANCE

Scientific data provenance spans the creation of data through their point of dissemination and beyond. It consists of a combination of static metadata (e.g., information on data sources, computing environment, versions of scripts) and dynamic metadata (e.g., operational steps and intermediate data values returned as a script executes). Static metadata can be stored as items in a database or file and accessed through traditional database queries or file examination (Chapter 13). Dynamic metadata require much more complex organization because they need to capture information about how data flow through a series of computations.

Dynamic metadata typically are organized using a structure called a provenance graph. A provenance graph is a mathematical graph consisting of nodes and edges, where the nodes represent operational steps or pieces of data and the edges represent how data flow as input into one operation or output from another operation or how control flows from one operational step to another.[9] In all the examples in this chapter, nodes are depicted as ovals or rectangles, and edges are depicted as arrows.[10] Nodes also are labeled with descriptive names.

Data provenance is an area of active research, and many challenges remain to be overcome.[11] For example: exactly what information should be collected as a script executes? Since the provenance graph may be very large and complex, what is the best way to store, query, and visualize it? If the provenance graph is too large for practical use (perhaps orders of magnitude larger than the original data), will it be necessary to limit the information it contains? The technical problems of how to capture and manage data provenance are formidable and are perhaps best addressed by computer scientists. But the problems of what information to capture and how to use it are best addressed by researchers who actually produce, use, and re-use the data. Close collaboration between these domain scientists and computer scientists will help to address these challenges, define others, and advance the field.

We believe that a first approach in such a close collaboration is to incorporate data provenance into existing tools that domain scientists already use.[12] In the remainder of this chapter we describe initial efforts to do this for the R statistical language. R is widely used by biologists and environmental scientists for data analysis and visualization and provides a good example of an integrated environment that supports both execution of pre-existing scripts and interactive console sessions.[13] We expect that once scientists have an opportunity to use data provenance, they will discover new and innovative uses for it that will inform the development of future software tools designed to capture and manage it.

COLLECTING DATA PROVENANCE IN R

Our approach to collecting data provenance in R involves two steps (fig. 14.1).

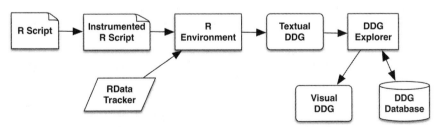

FIG. 14.1. Collecting data provenance in R. DDG = data derivation graph. See text for an explanation.

First, the scientist annotates a working R script by adding calls to functions contained in a special library, RDataTracker, to create what we call an instrumented R script.[14] These function calls cause RDataTracker to record data provenance as the script is run in an R environment (such as the R console or RStudio). When the script has completed its execution, RDataTracker saves the provenance graph (which we refer to as a data derivation graph or DDG) in text format.

Second, the scientist uses a separate stand-alone software tool, DDG Explorer, to read the text-format DDG and display it as a visual DDG. DDG Explorer also allows the scientist to store the DDG in a DDG database and to query the DDG in various ways. RDataTracker is written in R and designed to work specifically with R scripts, while DDG Explorer is written in Java and can read textual DDGs from any source.[15]

The amount of information captured in the DDG is a function of the number of annotations the researcher adds to her script. A minimal DDG can be generated by adding just a few lines of annotation, and by adding more annotations the researcher can collect additional data provenance.

In a few lines of code, we can illustrate a minimally annotated R script designed to capture data provenance while processing data from a weather station:

```
# Initialize the provenance graph
ddg.init("daily-solar-radiation.r", "ddg", enable
  .console=TRUE)
# Read the data from a file
data.file <- "met-daily.csv"
start.date <- "2012-01-01"
end.date <- "2012-03-31"
variable <- "slrt"
raw.data <- read.data(data.file, start.date,
  end.date, variable)
plot.data(raw.data, "R")
# Adjust for sensor calibration
calibration.parameters <- read.csv("par-cal.csv")
calibrated.data <- calibrate(raw.data, calibra-
  tion.parameters)
```

```
plot.data(calibrated.data, "C")
# Remove questionable values
quality.control.parameters <- read.csv("par-qc.csv")
quality.controlled.data <- quality
  .control(calibrated.data, quality.control
  .parameters)
plot.data(quality.controlled.data, "Q")
# Replace missing values
gap.fill.parameters <- read.csv("par-gf.csv")
gap.filled.data <- gap.fill(quality.controlled
  .data, gap.fill.parameters)
plot.data(gap.filled.data, "G")
write.result(gap.filled.data)
# Save the provenance graph
ddg.save()
```

The two DDG annotations are shown in boldface. The call to ddg.init(. . .) initializes the provenance graph and identifies the script to be executed (daily-solar-radiation.r), the folder where the data provenance will be stored (ddg), and whether console activity will be captured. The call to ddg.save() at the end saves the data provenance in a standard text file in the ddg folder, where it can be viewed by DDG Explorer.

An excerpt from the resulting DDG, as displayed in DDG Explorer (fig. 14.2) illustrates useful graphical conventions to display provenance. The ovals are operational nodes that represent the execution of R statements and are connected by solid arrows that indicate control flow. The rectangles are data nodes that represent pieces of data and are connected to operational nodes by dashed arrows that represent data flow. So, for example, the operational node that represents the R statement data. file <- "met-daily.csv" is connected by a data flow edge to the data node data.file, which in turn is connected by a data flow edge to the operational node raw.data <- read.data(data.file, start. date, end.date, variable). The DDG thus provides a graphical depiction of how the script was executed (i.e., an execution trace), including the order in which statements were executed, and how data flowed through the script.

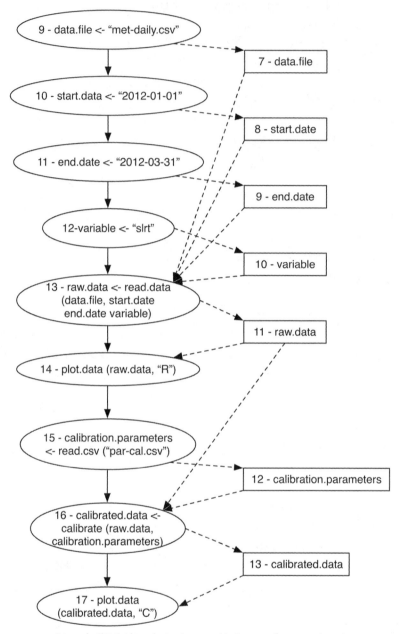

FIG. 14.2. Example DDG (data derivation graph). See text for an explanation.

The DDG also is interactive. By clicking on a data node in DDG Explorer, the researcher can see its value (e.g., clicking on data.file reveals that its current value is "met-daily.csv"). DDG Explorer shows more than single values, however. It also can display complex values such as entire data frames or image files created by the R plot command. The DDG also contains static information about the computing environment, including the platform, version of R, name of the R script, and date and time of execution.

MANAGING PROVENANCE VOLUME

The size of the DDG is a matter of real practical concern. Even the simple script shown earlier generates a DDG that is much larger than the script itself. For longer and more complex scripts, the resulting DDG might be huge if separate nodes were created for every item of data and every statement.

The size of a DDG can be managed by limiting the amount of information that is saved in it or that is displayed to the user. One way to limit the amount of information saved is through a user-defined setting that controls how much intermediate data should be saved. Alternatively, RDataTracker's console mechanism (which captures nearly every R statement) can be disabled and replaced with calls to ddg.function() or ddg.procedure() to mark specific points in the script where the researcher would like to create operational nodes in the DDG. The amount of information displayed to the user can be limited by introducing abstraction into the DDG through the addition of calls to ddg.start() and ddg.finish(). In DDG Explorer, any section of the DDG that was created by a pair of ddg.start() and ddg.finish() calls can be collapsed to a single node by clicking on either the start or the finish node or can be expanded to show the intervening details by clicking on the corresponding abstraction node (shown as an oval with blunt ends in fig. 14.3).

This feature is demonstrated with these lines of code:

```
ddg.start("Create contour plots of trees")
attach(trees)
plot.1 <- trees[plot==1, ]
plot.2 <- trees[plot==2, ]
```

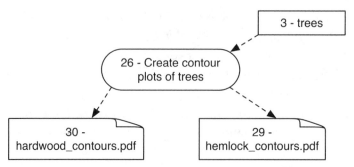

FIG. 14.3. Abstracted node hiding internal detail. See text for an explanation.

```
plot.3 <- trees[plot == 3, ]
ddg.procedure("extract tree location",
   ins=list("trees"), outs=list("plot.1", "plot.2",
   "plot.3"))
# . . . Many lines of the script omitted.
pdf(file='hemlock _ countors.pdf', width=4.75,
   height=6, colormodel='cmyk', pointsize=9)
# . . . More lines of the script omitted.
dev.off()
ddg.finish("Create contour plots of trees")
```

In this script, calls to ddg.start() and ddg.finish() create the abstraction node Create contour plots of trees in the resulting DDG. In DDG Explorer this node may be collapsed to show the abstracted view (fig. 14.3) or expanded to show the intervening nodes and edges (fig. 14.4). In the first of these (see fig. 14.3), all we see is that there is an operation to create a contour plot of the trees. This operation takes trees as input and produces two contour plots as output, but the details of how these contour plots are computed are not shown. Additional details can be highlighted with, for example, thicker borders (see fig. 14.4). This detailed view (see fig. 14.4) also contains still more abstracted nodes, which the user could expand to reveal still more detail. The example also shows how an operational node may be created and its inputs and outputs defined by a call to ddg.procedure().

Querying the DDG is another way to reduce the amount of information presented to the user all at once. DDG Explorer currently supports two types of queries that operate over a single DDG: what is the deriva-

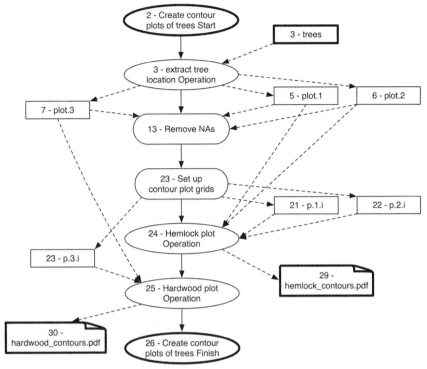

F I G . 1 4 . 4 . Node from fig. 14.3 expanded to show internal detail. See text for an explanation.

tion history of this datum; and how was this datum used in subsequent calculations? These queries have a number of practical applications, including verifying the value of a particular datum and tracing the impacts of a corrupted datum. By way of examples, we illustrate the subsequent uses of the datum dendro (fig. 14.5) and the result of querying the derivation history of hemlock-contours.pdf (fig. 14.6).

USING PROVENANCE TO AID IN SOFTWARE DEVELOPMENT

Data provenance was originally envisioned as a method for long-term documentation of published data, but we have also found it to be helpful during script development. For example, examining a DDG may be more helpful in determining whether a script does what it is supposed to do than simply looking at the end results of the script or scrutinizing the source code. The DDG also captures R error messages, which (if the error did not cause the script to terminate) may scroll by quickly in the R

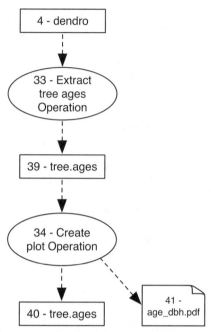

FIG. 14.5. Result of a query to determine what is computed from a value. See text for an explanation.

console without the researcher even seeing them. The DDG (fig. 14.7) shows the steps that were executed up to the point where an error occurred, together with the error message itself (retrievable by clicking on the error.msg node)

DDG Explorer also can be used for side-by-side comparisons of two R scripts. Lines are variously shaded to indicate modifications, additions, and deletions (fig. 14.8). Light gray indicates lines that appear in one file only; dark gray indicates the absence of those lines in the other file; and medium gray indicates lines that appear in both files but with modified contents.

USING PROVENANCE DURING DATA ANALYSIS

Scientific data analysis often involves a series of preparatory steps (e.g., reading, cleaning, combining, and extracting data) followed by analytic steps (e.g., visualizing and analyzing data, applying statistical tests, creating publication-quality plots). If any or all of these steps requires significant computational time, the scientist may wish to save

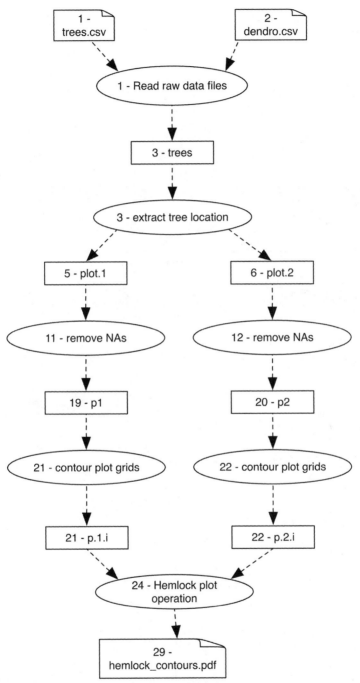

FIG. 14.6. Result of a query to determine how a value was computed. See text for an explanation.

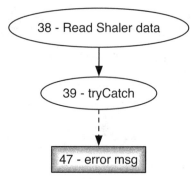

FIG. 14.7. Error message captured in the DDG (data derivation graph). See text for an explanation.

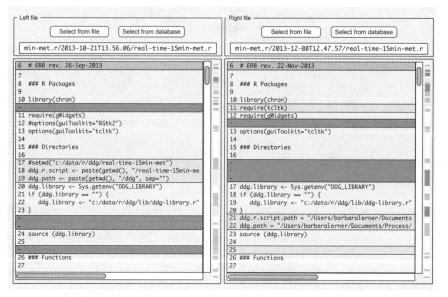

FIG. 14.8. Comparing two versions of an R script. DDG = data derivation graph. See text for an explanation.

the current system state at various points and restore it at a later point in the analysis.

RDataTracker supports this capability with the functions ddg.checkpoint() and ddg.restore(), which build on top of the existing R mechanism to save and restore a workspace but extend it in two key ways. First, when a checkpoint is restored, both the state of the R environment and the files that were current at the time of the check-point are restored. Second, the DDG maintains a complete history of all activity between the checkpoint and restore operations. When the re-searcher is viewing the DDG, this information is normally hidden by collapsing the intervening activity to a single abstracted node. However, if desired, the researcher can click on the checkpoint node and see all the intervening details.

Exploratory data analysis in R is often done by entering commands directly in the console. An important feature of RDataTracker is its abil-ity to collect provenance from console activity and connect that prove-nance to the information collected from running scripts. In this way, it is possible to keep a complete record of what the researcher has done, collecting provenance while the scientist works in the manner that best suits the problem at hand.

LESSONS LEARNED FROM R

The goal of the R provenance project is to learn what domain scientists would do if data provenance were brought to the tools that they rou-tinely use. One sign of a successful piece of software is that users not only use it as the developers intended but also use it in surprising new ways, pushing it into territory that the developers had not imagined.

Although RDataTracker and DDG Explorer are still under develop-ment, we have already been pushed by the domain scientists with whom we work to add uses of provenance we had not thought of. As computer scientists, we entered this project thinking the most important applica-tion of data provenance was long-term documentation of the process of data analysis and that its most important features were capturing data flow, control flow, and intermediate data values and controlling the vol-ume of provenance via abstraction and querying. Domain scientists ap-preciated these features but quickly looked for additional features, such as checkpoints and the ability to capture console activity and error mes-sages.

Conversely, some features we thought were important, such as completely automated provenance capture, appear to be less critical to domain scientists, who are willing to put some effort into provenance capture as long as they do not need to abandon their normal work rou-tine and adopt a new technology. RDataTracker and DDG Explorer are proving to be a fruitful middle ground, allowing computer scientists to explore new ideas about provenance while giving domain scientists provenance support within a familiar environment.

DISCUSSION

An essential first step in the reproduction or repetition of research is gaining an understanding of the original work, including its objectives and larger context as well as the origin and use of its data. Metadata are essential to gaining this understanding, and data provenance, in particular, adds reliability and trustworthiness to the data. The practical uses of data provenance, however, may change over time.

In the short term (weeks to months), the scientist who collected the data or is creating usable data sets may use data provenance to develop and troubleshoot scripts and to check the derivation and subsequent use of particular data values. In the near term (one to several years), other researchers may use data provenance to help them understand the original analysis, to examine the derivation and subsequent use of particular data values, to better use the data in further analyses, and to better re-use the analysis in other applications. Reproduction of the analysis using the same hardware and software still may be possible. In some cases, a scientist may wish to re-run the original analysis, particularly if there is reason to doubt the results or methods or if the results are especially important or controversial. In other cases, examination of the data provenance may be sufficient to generate understanding of, and trust in, the original analysis. We anticipate that this second use of data provenance will predominate with the passage of time since the original study.

Finally, in the long term (decades), reproduction of the original analysis using the same hardware and software is unlikely to be feasible. Hardware, operating systems, and application software are upgraded, and eventually it becomes impossible to even run the original script. Data provenance and the source code, however, will continue to support understanding of what was done and how it was done in the original analysis.

NOTES

1. Ellison et al. (2006).
2. In the discussion that follows we use the term "script" to refer to the software (often written or customized by the scientist) used to analyze and

transform scientific data. The journal *Science,* for example, requires that all data and computer code used to support results in a manuscript accepted for publication be available to all readers and that data sets be archived in a publically accessible database with appropriate citation (a digital object identifier [DOI]) in the published paper. See http://www.sciencemag.org/site/feature /contribinfo/prep/gen_info.xhtml#dataavail.

3. Boose et al. (2007).

4. *Oxford English Dictionary* online (http://www.oed.com/).

5. By this definition, data provenance is a kind of process metadata (Osterweil et al. 2005).

6. R: http://r-project.org/; Matlab: http://mathworks.com/products/matlab/ (an open-source alternative is Octave: http://gnu.org/software/octave); Python: http://python.org/.

7. Concurrency is a technique commonly used in interactive software and in high-performance software. A common type of error in concurrent software is a race condition, which occurs when two threads within the software share data and the outcome depends on the order in which each thread uses and modifies the data. In such cases, re-running the same software on the same input can produce different results.

8. Recall the distinction made in Chapter 1 between types and instances.

9. Technically, the provenance graph is a directed acyclic graph because its edges have direction and do not form any closed loops.

10. The software tools described in the following use different colors to indicate node and edge types; these are represented here using different line types and shades of gray.

11. Areas of recent research include developing methods for storing provenance graphs using a standardized notation (Ellkvist et al. 2009), connecting provenance graphs created with different tools (Lagoze 2013), associating provenance graphs with published data (Rüegg et al. 2014), visualizing provenance graphs (Horta et al. 2013), and using database technologies to query provenance graphs (Gadelha et al. 2011).

12. A number of scientific workflow systems, including Kepler (Bowers et al. 2012), LONI (Mackenzie-Graham et al. 2008), Taverna (Missier et al. 2008), Trident (Simmhan et al. 2008), and Vistrails (Koop et al. 2013), already support the collection of data provenance, but these specialized software tools often have a learning curve that domain scientists are simply unwilling to climb. In workflow systems, computational components are connected by data flow edges and provenance records which components execute and where their data come from and go to.

   Provenance is also collected by higher-level process languages such as Little-JIL (Lerner et al. 2011). In Little-JIL, computational steps are hierarchically decomposed into substeps connected by control flow edges; provenance records which steps are executed and what data passes between steps.

13. http://www.r-project.org (R Core Team 2015). R and Matlab, analytical tools widely used by domain scientists today, generally do not collect data provenance. However, there has been some recent progress in collecting provenance from R. CXXR (Runnalls and Silles 2012) is an implementation of the R interpreter that collects limited data provenance as an R script executes. R Markdown (Baumer et al. 2014) is a tool that runs inside RStudio and produces a human-readable version of an R script along with its inputs and outputs. The RDataTracker we discuss in this chapter (Lerner and Boose 2014b) is a library of R functions that collects provenance from the execution of R scripts and from R console sessions.

14. The RDataTracker library and DDG Explorer software and documentation can be downloaded from http://harvardforest.fas.harvard.edu:8080/exist /xquery/data.xq?id=hf091. For additional technical details, see Lerner and Boose (2014a).

15. RDataTracker and DDG Explorer have been developed with the aid of Shaylyn Adams, Nicole Hoffler, Antonia Miruna Oprescu, Luis Antonio Perez, and Sofiya Taskova. DDG Explorer incorporates the freely available software Prefuse (http://prefuse.org) for drawing graphs, Jena (http://sourceforge.net /projects/jena) for storing and querying DDGs, and iDiff (https://github.com /djromero/iDiff) for comparison of R scripts.

PART THREE INTEGRATION AND SYNTHESIS

## CHAPTER FIFTEEN

Turning Oranges into Apples

USING DETECTABILITY CORRECTION AND BIAS HEURISTICS TO COMPARE IMPERFECTLY REPEATED OBSERVATIONS

Morgan W. Tingley

It is often necessary to distinguish true state changes occurring through time from artifacts that reflect different methods. Within-sample replicates increase the precision of estimates, reduce bias, and permit the separation of observation and state processes from repeated observations.

Documenting long-term changes in biological systems requires empirical studies that span time frames from decades to centuries. Such time spans generally preclude planned experiments, but revisiting historical research programs or sites and repeating past methods or resurveying sites are being used to infer long-term changes. The unplanned nature of such resurveys, along with the uncontrolled environment, in which time becomes one of the treatments, results in imperfectly repeated samples. An important emerging issue is how researchers can use such imperfect repetitions—from which data may have been collected using different methods and for different purposes—to make comparisons that are statistically and inferentially robust.

This chapter reviews inherent problems of resurveys and summarizes methods that help account for imprecision and biases in methods in the design of resurveys and analysis of the resulting data. Many of the prescriptions presented also can be used to compare repeated measurements taken over short time spans (e.g., days, months, years), although such replicates often minimize bias by having been designed when the first sample was collected. Without such careful planning, however, methodological bias increases with the time elapsed between samples. Although repeated observations made a month apart might result in a comparison of apples to apples, more temporally distant observations may appear to be comparisons of apples to oranges.[1] But they are both fruits, and careful thought allows for meaningful comparisons between imperfectly repeated samples.

REVISITING TEDDY ROOSEVELT'S LIST OF BIRDS OF WASHINGTON, D.C.

In March of 1910, *Bird-Lore* magazine published "President Roosevelt's List of Birds," documenting the 93 species of birds observed by Theodore Roosevelt during his time in the White House from 1901 to 1910. The list was published by L. W. Maynard, who had persuaded the president to recount the list from memory, and it was delivered to Maynard—in the president's own handwriting—within 24 hours of the request.[2] The published list includes natural history remarks in addition to species names, indicating that, for example, Red-headed Woodpeckers, Eastern Screech-Owls, and Orchard Orioles all nested on the White House grounds, whereas Ruffed Grouse (undoubtedly a game bird prized by President Roosevelt) once had been seen nearby on Rock Creek (now the site of the National Zoo).

Much has changed in Washington, D.C., since 1910, and historical surveys—such as Roosevelt's list of birds—have the potential to provide us with data about how both individual species and entire species assemblages have responded to environmental change over time. Urbanization, for example, likely has eliminated many of Roosevelt's White House–breeding birds from Washington, particularly fragmentation-sensitive species like the Wood Thrush. At the same time, Roosevelt's list omits several species that are now very common in Washington, D.C., such as the Red-bellied Woodpecker, which was first recorded nesting in Maryland in 1903 and has steadily expanded its breeding range north during the twentieth century.[3]

At the same time, any formal comparison between Roosevelt's list and contemporary survey data—that is, a modern repeated sample or resurvey—faces enormous statistical and inferential challenges. Maynard appends to her published list Roosevelt's disclaimer: "Doubtless this list is incomplete; I have seen others that I have forgotten" (p. 55). Roosevelt himself, although by all accounts a devoted scholar of natural history, had many other things besides bird-watching competing for his attention and time while he was president. We have, for example, no record of the time he spent wandering the White House grounds, no spatial extent of his observations away from the White House, only limited evidence of the seasonality of his observations (e.g., anecdotal observations of "nesting" or "winter"), infrequent qualifiers of abundance (e.g., "one pair" or "very abundant"), and no record of his methods (did he know the bird songs, use binoculars, or have a regular walking route?).

Relative to contemporary systematic bird surveys, such as point counts or line transects (see Chapters 6–8), comparing Roosevelt's bird list to modern data is like comparing apples to oranges. A direct (naïve) comparison poses numerous opportunities for methodological, sampling, and inferential bias to obscure or otherwise misrepresent apparent differences over time. Nevertheless, historical data, surveys, and experiments hold incredible value as temporal baselines for the modern ecologist.[4] More generally, in the many fields discussed in this book, including historical geography (Chapter 3); observational ecology, evolution, marine biology, and biodiversity (Chapters 4–8); experimental biology and biomedical science (Chapters 9–12); and animal physiology (Chapter 16), repeated measurements taken at different times (which for convenience I refer to throughout this chapter as "resurveys") may be compared not for the purpose of generalizing among all replicates (see Chapters 9, 10) but rather to describe the temporal change itself. Whereas not every historical data set, including Roosevelt's White House list, is ripe for a new sample or resurvey, many data sets that can be repeated only imperfectly present opportunities for strong inference.[5] The challenge to the contemporary scientist is to foresee these challenges and take precautionary measures—in planning the resurvey, collecting the data, checking and cleaning the data (Chapters 13, 14), and analyzing data—both to minimize bias and, when possible, to separate any biases from underlying signals of systematic change.

PLANNING A RESURVEY

Repeating a previous study—conducting a resurvey—is a very particular type of scientific endeavor that involves an extensive planning process. As part of this planning, there are many opportunities for scientists to make methodological choices that will either help or handicap subsequent analyses, comparisons, and inferences. The key planning questions are these: Which variables will we match during the resurvey, and which ones won't we match? And how identical will our repeated samples be, or how imperfect must they be?

The answers to these questions depend on the motivation(s) for the resurvey. Temporal resurveys are generally designed to capture some specific aspect of temporal change, such as changes in population structure, pollutant loads, distribution and emergence of infectious diseases, land use, or climate. If our goals are to understand impacts due to climatic change, the choices we make in how to structure a resurvey will be different from those we make for a resurvey with goals to understand the evolution of a novel virus. In virtually any case, however, samples taken repeatedly through time are unlikely to be affected only by a single driving agent of change.

Very little in nature exhibits true stasis. A single plot may be the site of an experiment sampled every 40 years, but during any 40-year interval, the research site may have been surrounded by urban growth; subject to increasing, localized ozone and particulate pollution; impacted by acid rain; or exposed to novel temperature or precipitation regimes due to regional climate change. Time itself can be its own driving force; even cells living in controlled environments develop new, unforced cycles and long-term shifts in circadian rhythms.[6] For any scientist planning a resurvey, all of these factors can be viewed as either sources of bias or sources of inference. Robust inference on one factor, however, requires excluding or statistically controlling for the others.

The critical choice involves which conditions or methods to repeat and how precisely to repeat them. Undoubtedly, if one can methodologically control for, say, climatic change in a study of urbanization impacts, it would be advantageous to do so. Often, however, the choices available when planning a replicate are not about controlling for particular factors but about making specific choices about where to sample, what to sample, and how to sample. Comparisons between repeated samples are most straightforward when the original conditions and methods can be repeated precisely at each subsequent sampling time. This can create tension, however, between the need to match original methods and the desire to use the best available, and often the newest, methods.

Precisely repeating methods in resurveys may not always be possible, and even when it is, it may not be practical. In an extreme, but perhaps not unreasonable, example, such precision can approach the surreal. Imagine a scientist surveying bird populations in 2017 re-using methods identical to those used in 1915. As she approaches the beginning of her survey route, she pulls out her pocket watch and notes the time in her field notebook, pausing briefly to refill her fountain pen. Although she normally wears contact lenses, for this resurvey she has left them at home in lieu of her grandfather's wire-frame glasses. These glasses don't correct her astigmatism, but the prescription is close enough. As the survey begins, she alternates between writing down the species she detects and peering through a pair of Zeiss binoculars manufactured in 1904 that she purchased from an online antique's dealer. Although she has been working in the area for a while, every so often she hears a bird she is not quite familiar with. Not allowed to use the collection of recorded bird sounds on the smartphone she left at home, she stops her census and attempts to find the singing bird. If she's lucky enough to find the bird but doesn't recognize it, without field guides for the area, she takes copious notes, draws a quick sketch, and hopes that she can match them to museum collections or reference books the following week. Alternatively, she can come back to the site with a shotgun.

There is little advantage in deliberately handicapping a researcher doing a resurvey, but there is no denying that an ornithologist working in 2017 would have distinct advantages in her ability to detect and identify birds over her colleagues working in 1915, even if the assemblage being surveyed was identical. The way forward lies in distinguishing sampling methods that affect the ability to detect or observe the same state or process that occurred between the surveys from methods that risk having the resurvey sample a fundamentally different state or process. Both can strongly bias inference, but it is possible to account statistically or methodologically only for the former.

The Importance of Location

When choices must be made that can result in sampling different states or processes, location is of paramount importance. In many experiments seeking to gain generality through replication or repetition, the same methods are repeated at many locations to reveal how processes change across space. When inference is being made across time, however, robust inference requires that repeated samples be collected in the same place. The strength of a design for a resurvey lies in the assumption that true changes to the system arise only from processes that themselves are changing through time.

In practice, resurveying at the same location poses real difficulties and involves many types of uncertainty that implicitly are often ignored.[7] For example, a written description of a location (e.g., "1 kilometer south of Mirror Lake on marked trail") can be interpreted many ways; it assumes that relative measurements are precise (Was it exactly 1 km, not 1.5 km? As the crow flies or trail distance?); that place names are known perfectly (Mirror Lake in Wallowa County, Oregon, or Mirror Lake in Clackamas County, Oregon?); that landmarks do not change over time (Has the trail been re-routed?); and that original sampling was done at a single locality centered on the described point. Even when satellite-determined geographic coordinates are used, remnant uncertainty can still result in downstream bias in inference.[8] Current technology allows field locations to be identified accurately in space with sub-meter precision, but the concept of a sampling locality is discipline-specific, organism-specific, and question-specific, and thus the question "Just how close should replicate localities be matched?" requires context-specific definition (see also Chapters 8, 16).

Other characteristics of resurveys, such as date and time, also are important to match with prior samples. For example, it might make little sense to repeat in February earlier fungal surveys conducted in October. Similarly, repetitive sampling of insect traps left open for 48 hours should be started at the same time of day or after the same number of degree days have elapsed. Challenges appear when temporal characteristics of a sample are related to non-Gregorian calendars. For example, repetitive sampling of marine invertebrates in intertidal zones should be timed to the lunar cycle. Knowing the relevant temporal cycles for sampling is part of good scientific practice, but it can be complicated when cycles change over time (see Chapter 16). How should a study of changing wildflower or coral abundance control for changes in phenology resulting from a warming climate? Such interactions can change inferences inadvertently, requiring additional attention to sampling details.

Although place and time deserve important consideration when planning repeated observations, many other aspects of methodology should be changed with forethought. As in the previous example of the bird surveyor, it makes little sense to deliberately handicap oneself. Samples taken at any point in time not only are comparisons to the past but also provide baselines for future resurveys (see Chapters 6–8). If new or changed methodologies are expected to change strongly the relationships between observations and underlying processes, a sub-sample of measurements should be taken that use both methods (the original and the new one), which allow for correlation and comparisons among results from both. If methods (e.g., using a field guide and new binoculars) are expected to change only the relative precision or accuracy of the observational method, subsequent analytical tools can be used to calibrate the comparisons, especially if temporal replicates (i.e., within sampling bouts; see Chapters 6–8) are collected.

Preparing Data for Comparison

Once resurvey data have been collected, numerous operations can be used to increase the robustness of inferences and conclusions drawn from comparisons between and among the surveys. Two broad classes of operations are characterized by filtering and standardizing the data.

Filtering involves carefully selecting which data are used for comparisons across surveys. Unlike the removal of outliers, filtering for comparison excludes data based not on their values but rather on their metadata.[9] Data that were observed or collected under conditions that are unusual or extreme with respect to the rest of the data can be excluded from the analysis to limit overall heterogeneity in temporally repeated samples.

The types of metadata that may be important to filter will differ from one study system to another. Common characteristics of data that may be important to remove by metadata filtering include these:

Location. The accuracy or precision of location may vary from site to site or sample to sample. Filtering can exclude sites based on the magnitude of uncertainty of one site relative to others.

Time. Samples will be collected at different times of day and times of year. It may be advantageous to eliminate samples collected unusually early or late (e.g., late in the day or year).

Weather. If weather or other atmospheric conditions can affect sample data and if the weather conditions of sampling are known or can be reconstructed, data collected during extreme weather events can be removed prior to analysis.

Observer. In many types of data collection, multiple individuals are responsible for collecting data, and their identities are usually associated with these data when archived. In retrospect, however, not all observers may be found to have been equally skilled at data collection. If only a small fraction of the total data set was collected by one or a few less skilled observers, the data that they collected should be evaluated carefully and, if necessary, filtered from the analyses.[10]

Observed units. In both initial surveys and resurveys in which data on every individual (e.g., species, cell, genome) encountered or sequenced are recorded, it is common to limit analysis to an a priori group within a specific level of biological organization relevant to observation. Often this may be done because only certain groups (e.g., particular groups of species, particular genes or gene families) have enough data to warrant further analysis or because the data collection methods were specific to individual entities with certain characteristics (e.g., mammal traps generally catch species within certain size limits).[11] In many resurveys over time, however, there is the additional complexity of taxonomic and other nomenclatural changes. For example, a species name recorded for the first replicate may now be ambiguous taxonomically because the species has been split into two sympatric taxa. Serogroups of pathogens (see Chapters 11, 12) may evolve, change, or be renamed. Some names of species (or other entities) also may be ambiguous due to changes in language and usage, particularly if common or informal names are used (see Chapter 4). If codes are used during data entry and no full look-up table of complete names is provided, there may be certain data that will warrant exclusion from comparative analyses.

Collection method. In cases in which multiple methods may have been used to collect data, particularly if there was a primary, commonly used method and a secondary, infrequently used one, it may be useful to work with the data from only the commonly used method.

In all cases, however, filtering data by these (or any other) characteristics depends on specific choices made by the researchers involved. Such choices should be reported in analytical methods, and filtered (or flagged) data sets should be provided alongside the originals (see Chapters 13, 14). Filtering can help match many of the underlying conditions that govern how data were collected in resurveys. In contrast, standardization of data is used to reduce heterogeneity across repeated samples and to bring their measurements onto comparable scales.

Standardization. This is especially important when data from resurveys have a higher information content than previously collected data. Because newer surveys also serve as baselines for the future and because methodological advances frequently increase the efficiency of data collection, resurveys generally collect more data and more types of data than older ones. For example, an observational survey for which only data on the presence or absence of a particular species were collected as a baseline may count numbers of individuals (abundance) during a resurvey. Timed observations previously collected once each minute or hour may now be collected once per second or millisecond. Genetic studies that once sequenced only a single protein such as Cytochrome b may be repeated by sequencing the full mitochondrial genome. In some of these studies, such changes manifest as an increase in data density, but in others, the very nature of the data themselves also changes.

When comparing such data sets, both need to be placed on common scales. Usually this means that the data with the higher information content (generally, but not always, the newer data) need to be aggregated to a lower information content. For example, abundance data can be recoded as presence (abundance > 0 = presence) or absence, measurements taken every millisecond can be averaged over minutes, or millions of genetic variants can be aggregated into population groups based on geographic isolation or linguistic divisions.

Depending on how methodologies have changed over time, comparisons of values may be best standardized by comparing ranks.[12] Such a conversion can be critical when there is no clear way to numerically standardize data collected by different methodologies. Examples include when technology has changed to the extent that data are measured on inconvertible scales or when inference on relative abundance is desired but resurveys used two incompatible ways of counting individuals. In these cases and many others, comparisons of ranks across resurveys can be used for inference.

STATISTICAL COMPARISONS OF DATA FROM RESURVEYS

Once the data have been filtered and standardized to reduce potential bias and other confounding factors, the researcher can proceed to analysis and inference. The wide range of statistical tests that can be used to compare data from surveys and resurveys is far too large to cover within a single chapter, but the choice of statistical test rarely is the stumbling block for most researchers. Rather, the difficulty lies in using statistical methods that account for as many differences and imperfections as possible in resurvey data. In other words, while it is easy to test for differences between two (or more) repeated samples, it can be difficult to draw appropriate inferences from those tests when samples are repeated imperfectly.[13]

There are two common strategies that researchers use in such situations. One bypasses the imperfections altogether, whereas the other deals with them explicitly. The first method for statistically comparing imperfectly repeated samples can be described as "check and carry on."[14] The motivation for check and carry on is to bypass the unpleasantness of imperfections in the repeated data by first testing for effects on measured variables of known biases in (for example) sampling methods. If such testing reveals no significant effects of the bias, no further barriers remain for robust analysis and inference.

To illustrate check and carry on, we return to an ornithological example. Two surveys of birds were done nearly 50 years apart on two mountainsides in Papua New Guinea.[15] The goal of the resurvey was to test for climatically driven uphill shifts in species distributions. A concern of the resurvey was that increases in observer effort could bias estimates of species distributions: more effort (time spent surveying) would lead to more detections, resulting in apparent range shifts detected because of increased effort, not real range shifts that occurred as a result of climatic change. To reduce the probability of drawing this incorrect conclusion, the researchers asked whether apparent range expansions (either upslope or downslope) had occurred more frequently than range contractions.[16] Their logic was that if observed range shifts were primarily the result of increased effort, expansions (the probabilistic outcome of increasing observations) should outnumber contractions. Instead, they found no expansion-biased trend. Having determined that effort did not bias the resurvey in an expected way, they proceeded with the analysis as if the resurvey used methods identical to those used in the initial survey.

Check and carry on has obvious appeal. It is easy to understand and implement. It justifies the use of standard and familiar statistical methods. It appears logical, and results are easy to communicate. Because the researcher tests an alternative hypothesis that could explain differences between repeated samples, biases that could have a major influence on the conclusions should be identified via this test.

But this apparent appeal can mask important weaknesses. In most implementations of check and carry on, the statistical tests used to detect biases in methods are overly liberal. Generally, such tests for bias are evaluated as one would a null hypothesis (i.e., there is a bias present only if the null hypothesis is rejected with a P value less than 0.05; see Chapters 9, 10). This test ignores whether there is actually enough power even to detect the bias, so it is informative only in those cases in which the bias is so strong or pervasive that its impact is undeniable.

Tests for check and carry on would be more powerful if bias were assumed to be present and the P value expressed the probability that bias, in fact, occurred. Although this approach would increase our confidence in the conclusions of check and carry on, it suffers from the alternative option it leaves the researcher: "Carry on or . . . ?" Because check and carry on leaves researchers with no alternatives if bias is found, the approach leads to perverse actions. Either the researcher throws out all of the research because the repeated samples are biased or she keeps hunting until she finds a statistic that shows that the bias has no effect.

We know that some amount of bias is pervasive in resurveys. An alternative to check and carry on is to include the bias in the analysis but use the strength of data to differentiate among the sought-for signal, random noise, and methodological biases. Most researchers appreciate that environmental gradients and individual differences can affect results and design their observational programs or experiments accordingly. Hierarchical statistical models incorporate such factors (often called "nuisance parameters") alongside the driving variables of inferential interest.

In particular, hierarchical models begin with the assumption that parts of the sampling process that make repeated samples imperfect or imprecise are themselves reflecting random processes. For example, two birders may differ in their ability to detect particular species, so richness estimates at a given site will be affected by the composition of species present. In this case, not only should species richness be modeled as a function of the environment but the observational process (differences among observers) that gave rise to estimates of species richness should also be modeled.

Such divisions of observation processes and the stochastic outcomes of observations lend themselves to a class of hierarchical models called state-space models (alternatively called hidden Markov models). The logic behind state-space models is that data are representations of an unknown (true but unobservable) state seen or measured through the observation or data collection process. A few examples should convince. A measurement of how much of the ground is shaded by the canopy of a tree can be considered an index of the true but unobservable value of canopy cover. Because some mammals frequently are not seen, the number of species counted is an underestimate of the true number of species present. The count of cancer cells in a blood sample is not equivalent to the prevalence of cancer in a patient's body. And contour lines drawn on a map are only a representation of the actual changes of elevation on the ground.

In fact, most scientific measurements can be viewed as imprecise representations of a true, unobserved state that results from imperfect observation processes. Most of the time, however, the imperfection of the observation process is assumed to be part of the error that is modeled in any statistical test. It is only for particular cases, such as the comparison of imperfect repeated observations in which the observation process differs across repetitions, that it becomes advantageous to model the observation process directly using state-space models.

How does a state-space model work? The simplest state-space model is composed of two distinct but linked processes that describe how an empirical truth is investigated (fig. 15.1). The first process, which can be mathematically represented by an equation, describes the true (but unknown) state of the system as a function of various factors that are expected to influence it. The second process and its associated equation describe the observation process. The observations depend on the true state, but they also depend on various observational biases, such as the observer, the weather, or the time of day (thus the two equations are hierarchically linked). These biases are the very characteristics that make our resurveys imperfect.

OCCUPANCY MODELING AS AN EXAMPLE OF A STATE-SPACE MODEL

Occupancy models are a type of state-space model that are widely used for examining data on the presence or absence of species.[17] The true state of a measurement of species occurrence is whether a species is present. For most species, however, the ability to detect them is imperfect, resulting in false absences when the species is present but undetected or sometimes in false presences when the species is absent but is mistakenly detected (e.g., it is confused with a similar-looking or -sounding species). Such false absences and false presences abound in occurrence data in ecology and have been shown to strongly affect the conclusions about what factors influence the true state of occurrence.[18] Similarly, false positives and false negatives are common outcomes of testing for diseases and affect the conclusions about overall disease diagnosis and prevalence (see Chapter 10) as well as decisions as to whether one should undergo certain courses of medical treatments (see Chapter 12).

Occupancy models can be used to differentiate the true state (e.g., present or absent; sick or well) from the observed state (e.g., detected or undetected; positive test result or negative test result) when you only have data on the observed state. But occupancy models also need additional information on the probability of detecting the observed state: that is, the probability of actually detecting that a species is there or the probability of a true positive result from a diagnostic test. Sampling must also be conducted in a wide enough variety of conditions to quantify these probabilities. Examples include sampling in different weather

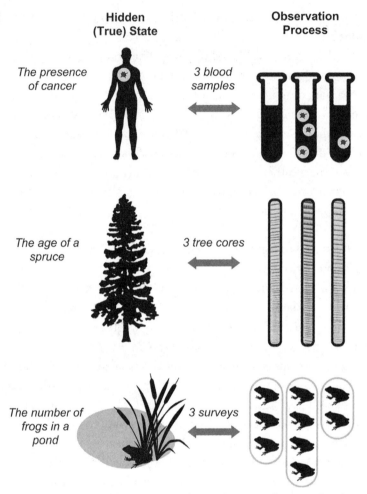

FIG. 15.1. State-space models representing how unknown truths are explored through an observation process. Although there is no requirement that more than one observation or sample be taken for each unknown state, in science we often repeat our observations. Thus, state-space models provide a bridge to understanding the true state while accounting for errors and biases that may exist in our observations.

conditions when it is expected that detectability will change or at different times of day when it is expected that blood pressure rises or falls. These within-sample replicates yield estimates of the probability of how the observational process varies within a survey when it is assumed that the overriding state (presence of a species, occurrence of hypertension) otherwise is not changing.[19]

Occupancy models are just one example of state-space models that use within-sample replicates to improve statistical inferences about processes of interest operating across individuals, samples, and surveys. Within-sample replicates account for both differences among individuals and biases caused by observation processes that could otherwise obscure differences and changes between surveys.

There are many occasions, however, when within-sample replicates are not available; Roosevelt's list of birds is but one of many examples. State-space models can be constructed without within-sample replication, but they are much less powerful than models that include within-sample replicates. Even if within-sample replicates are collected at only one point in time, however, they still provide some ability to correct for observational biases across the entire analysis, thus strengthening the comparison. Researchers not planning to compare data to historical baselines can use within-sample replicates to estimate current observational biases while providing better baselines for future studies.

LESSONS FOR FRUITFUL COMPARISONS

Many repeated samples are separated by long periods of time during which methods, data collectors, and observational conditions have changed. The fundamental problem with comparing these imperfect samples is that this accumulation of differences can bias inference on true, underlying change. Even the most carefully planned resurvey can suffer from the effects of time on the ability of researchers to detect change. Comparisons that appear to be of apples and apples may actually be comparisons of apples and oranges. Although I risk straining the metaphor, I should point out that both are fruits; reliable inference of change may require peeling away the orange-ish rind or coring the apple. To complete this apparent recipe, a five-step program can be used to turn oranges into apples (or vice versa):

Know whether you have an apple or an orange. The key to planning what to do with imperfectly repeated samples is knowing your starting point. This is particularly true for historical data, where initial evaluations of data and associated metadata may be incomplete. For example, a data sheet may have some vague locational data that are made specific in separate notes hidden away in a file drawer. Knowing exactly how much information you have allows you to identify, before starting the resurvey, the strengths and weaknesses in the data, and the potential biases for any future comparison. Knowing is the first step to solving.

Pick an apple that looks most like an orange. When planning a resurvey, knowing your baseline allows you to match conditions, idiosyncrasies, and methods that will allow for the best possible comparisons. It rarely makes sense to ask data collectors to use outdated, imprecise, or inaccurate methods and tools, but one should match, to the extent possible and reasonable, current sampling methods to those used historically.

Use a good metric for comparing apples to oranges. Despite idiomatic sayings, there are actually a number of good ways to compare an apple to an orange. For example, one could compare the diameters, spectral reflectance, sugar content, or genomes of these two fruits.[20] Filtered and standardized data facilitate robust comparisons across imperfectly repeated surveys.

Quantify and incorporate apple-ness and orange-ness. No matter the extent to which the previous steps are taken, an apple will always be an apple, and an orange will be an orange. Rather than trying to hide the differences, seek to understand them. Once you understand the differences, a better comparison can be made because you can account for those differences that will be present in any comparison of apples and oranges. By modeling the data collection and the observation processes, you can account for their contributions and focus on the remaining differences that are the signals of interest.

The more apples and oranges, the better. If the goal is to define apple-ness and orange-ness, one sample of each is insufficient. Within-sample replicates permit the separation of the observation process from the state process.

Regardless of how many steps are taken, the key message is to not pretend that apples are always being compared to apples. Identical replicates and precise resurveys rarely, if ever, occur. Most of the time, minimal differences among repeated samples or biases in measurements are assumed to be small. This assumption is likely to hold if methods are standardized and samples are collected carefully and in close succession. As sampling intervals increase, differences are magnified. At some discipline-specific point, it is hard to assume that even standardized protocols yield directly comparable resurveys, and then there is no denying that every apple is also an orange.

NOTES

The ideas present in this chapter emerged over many years of discussions with friends and mentors working on the Grinnell Resurvey Project at UC Berkeley, including Steven Beissinger, Craig Moritz, Michelle Koo, Jim Patton, Carla Cicero, Chris Conroy, John Perrine, Karen Rowe, and Kevin Rowe.

1. But see Chapter 16 for an apples-to-apples comparison that is achieved by increasing the time span between observations.
2. Maynard (1910).
3. Shackelford et al. (2000).
4. The past 20 years of ecology has shown a tremendous increase in the number of issues that are lumped as "global change." Whether climate change, land-use change, or change caused by invasive species, the world is changing at a global scale and every year sees ecosystems responding in myriad ways. Change in this new, highly dynamic era (often referred to as the "Anthropocene") is hard to measure in only yearly increments but is easily detectable in decadal or semi-centennial increments. It is to this purpose—providing long-term baselines—that historical data are showing their amazing value, whether locked in museums, libraries, or file cabinets.
5. Here, I use "historical" to broadly mean the older or original of any temporal set of replicates and do not imply any condition for what qualifies as historical. The time span necessary to understand change when comparing replicates is unique to the type of change and the study system. A time span of one year may be too long to understand *Drosophila* population dynamics but too short to measure tree-line migration.
6. Kaeffer and Pardini (2005); Borgs et al. (2009); O'Neill and Feeney (2014).
7. See Shavit and Griesemer (2009) for detailed discussion of locality problems in surveys and resurveys of biological diversity.
8. Rowe and Lidgard (2009).
9. Metadata are defined as data about data. See Chapters 13 and 14 for detailed discussion.
10. It may be worth commenting here more fully on what characteristics could be used to justify a priori exclusion of data from a particular observer. I do not mean to suggest that the data values collected by an observer should be evaluated as to whether the data values seem "right" or not. Rather, I recommend evaluating other characteristics of the type or quality of data collected. If the original data sheets are available, they can be checked for symptoms (e.g., writing something and scratching it out, excessive question marks, and so on) that may indicate that an observer was under-trained or not adhering to protocol. Making such a judgment is not trivial. In many disciplines it may not be possible or advisable to exclude specific observers. A detailed case study of observer skill and bias in field sampling is provided by Fitzpatrick et al. (2009).
11. Recent examples of such data filtering and aggregating of survey data to address specific questions include Gurdasani et al. (2015) for human

genetic diversity and genomics and Engemann et al. (2015) for biodiversity in Ecuador.

12. Data are re-coded to ranks first by sorting them (e.g., highest to lowest) and then by replacing the data value with its rank (e.g., highest value=1, lowest value=100). Ties (data with the same value) are assigned the same rank. Statistical analysis of ranks uses "nonparametric" or "distribution-free" methods because there is no explicit statistical distribution fit to the ranks. See Hollander et al. (2014) for a full treatment of nonparametric statistics.

13. Simple *t*-tests or repeated-measures analysis of variance are a common starting point (Gotelli and Ellison 2012).

14. Although this term sounds like the options for stowing luggage before plane travel, it derives from an attempt to shorten the expression "check for differences and then carry on."

15. The original surveys were done in 1965 (Diamond 1972) and 1969 (Diamond and Lecroy 1979). The resurveys were done in 2012 and 2013 by Freeman and Class Freeman (2014).

16. This type of erroneous inference is a Type I statistical error. See Chapters 9 and 10 for additional discussion of statistical errors.

17. A general reference for occupancy modeling is MacKenzie et al. (2006).

18. See, for example, Guillera-Arroita et al. (2014).

19. For example, two researchers walk through a section of a forest counting lady-slipper orchids (*Cypripedium acaule*) along fixed transects, and they count and flag 161 plants. Is this all there are? Either every orchid that is present has been accounted for or some were undetected. The researchers could walk the transect again the next day but at a different time of day when the light was different. It is unlikely that new orchids could have grown up and flowered within 24 hours, so a second sample would allow the researchers to calculate the probability that they detected a single orchid. The more within-sample replicates, the better the estimation of detection probabilities and the more accurate the estimation of occupancy (occurrence) of orchids. See Chapter 14 of Gotelli and Ellison (2012) for a complete elucidation of this example.

20. Scott Sandford ground up and compared samples of an apple and an orange using mass spectrometry. In a startling revelation (Sandford 1995), he concluded that, based on this method, apples and oranges are actually quite similar.

## CHAPTER SIXTEEN

Dissecting and Reconstructing Time and Space for Replicable Biological Research

Barbara Helm and Ayelet Shavit

Precise locations from global positioning systems and exact times from atomic clocks are often insufficient and potentially misleading attributes of an organism's movements or physiology. A close look at autocorrelation of repeated samples can indicate the potential for successful replication and help us to identify relevant patterns in space and time. Appreciating the many components of space and time and accounting for their combined effects on data will improve the replicability of observational and experimental studies.

"Time" and "space" are two fundamental dimensions in the characterization of processes and objects in the life sciences, and therefore, clear and unambiguous temporal and spatial information is a prerequisite to biologically meaningful replication. Given the availability of high-precision atomic clocks and global positioning systems, it may seem straightforward to return to the same point in space and to identify an equivalent point in time for the sake of observational or experimental repetition. However, even with high-precision technology, measurements and descriptions of time and space are neither accurate nor unambiguous for repetition and subsequent generalization. Consequently, repeating a laboratory manipulation or field survey at an approximately equivalent location in time or space can be a cause for additional variation and error, sometimes with substantial impacts.

In this chapter we draw attention to the importance of careful consideration of time and space for meaningful replication. As a first step, we show that seemingly straightforward assumptions about variation over time and space do not hold. Instead, the many aspects that contribute to an organism's time and space lead to substantial ambiguity and often incompatible practice. In order to tackle this ubiquitous problem, we view the operational meaning of "time" and "space" from three different perspectives—exogenous, endogenous, and interactionist—which we argue should not be conflated. We then illustrate the implications by presenting in a single table a wide range of paradigmatic examples grouped by spatiotemporal descriptors and reviewed sequentially from all three perspectives. Given the fundamental role of spatiotemporal information for repeating biological observations or experiments and its ubiquitous ambiguity, we conclude that such a reconstruction of "time" and "space" within a general, pluralistic, and multi-faceted table is not only new but also a first important step to be further developed. After all, our aim is to encourage further thought about appropriate measurements of time and space. We are convinced that replication of studies can gain substantially from recognizing and teasing apart different aspects of time and space instead of using a single universally standardized concept.

AMBIGUITIES OF TIME AND SPACE: PATTERNS OF VARIATION

Variation over time and space affects replication in many ways. The title of this book refers to a common interpretation of Heraclitus's (allegorical) insight that one can never step twice in the same river. One would tend to assume, however, that stepping into the river twice within very little time or space would make the events relatively similar. Generally, the closer in time or space that two sampling events occur, the more similar, and thereby the more repeatable, they will be expected to be. This expectation holds for some, but not all, biological patterns and processes (for useful contrasts, see Chapters 6–8 and 15).

The change in similarity between observations at increasing temporal or spatial distance (i.e., the temporal or spatial "lag") can be quantified by an autocorrelation function (ACF; Fig. 16.1).[1] The different forms of ACFs observed in biological data introduce some of the difficulties associated with replication. Importantly, there is wide variation in the form and distribution of the ACF, with substantial implications for repeatability and experimental design. The most extreme situation is no autocorrelation over time and space (see fig. 16.1, *panel A*), so that subsequent samples are statistically independent, implying that their similarity does not depend on their measured positions in time or space. However, some autocorrelation is commonly detected. In this chapter we emphasize ambiguities in time because ambiguities in space are discussed in general in Chapter 15 and elsewhere in greater detail.[2]

Some biological processes proceed directionally through time. In these instances, time is seen as continuous and directional, starting from an arbitrary point and proceeding through successive (and cumulative) elapsed time units (e.g., seconds or years). Biological processes that change directionally include the development and aging of organisms and stages of succession toward climax community. In such cases, repeated samples will become progressively more dissimilar to each other and the ACF may decay through time. Accounting for the change in the ACF in sampling or experimental protocols or by post-hoc analytical techniques would seem straightforward. However, the rate of decay may vary (see fig. 16.1, *panel B* and *panel C*), for example, between sites or individuals, and the factors that shape specific trajectories, for example, those of development and aging, are not completely known to researchers. Therefore, introducing a common correction factor can in fact introduce artifacts and reduce repeatability.

Differences in trajectories of development and aging are so relevant that researchers have introduced definitions of time that are alternatives to the passage counted in seconds or years. For example, data collected from growing individuals are often reported relative to sequential stages, without explicit reference to the passage of time.[3] Furthermore, growth rates of many organisms, for example, fish, are greatly influenced by ambient temperature, which often differs between studies and replicates.[4] In yet another convention, therefore, the time taken for growth is sometimes reported relative to cumulative ambient temperature, such as degree days (i.e., integrals of a function of time and concurrent ambient temperature). Similarly, in agricultural and ecological studies, the

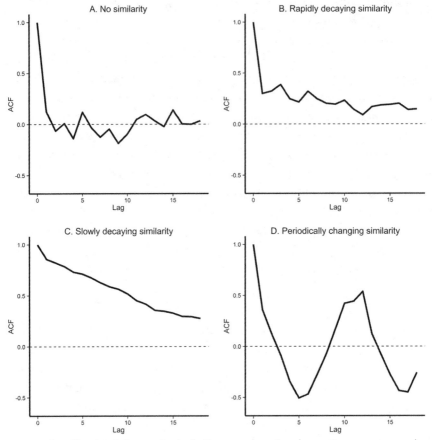

FIG. 16.1. Simulated changes in similarity over successive observations, as measured by an autocorrelation function (ACF). Simulated data and calculations were generated using the ACF in the stats library of the R software package (R Core Team 2015), with the kind support of Iain Malzer. The graphs show for four scenarios how the similarity of sequential observations declines with sampling intervals ("lags") over time and space. The ACF assumes values between 0 (no similarity), indicated by the dashed horizontal line, and ±1 (complete similarity); note that for lag=0 the ACF is always 1. *Panel A:* No similarity (successive observations are independent of one another). *Panel B:* Rapidly decaying similarity. Some similarity, but ACF decays rapidly over sampling intervals. *Panel C:* Slowly decaying similarity. Initially high similarity, with ACF decaying slowly through time. *Panel D:* Periodically changing similarity. Periodic processes may show a rapid decline in ACF followed by a subsequent rise. Here, the time between peaks is 12 observations, potentially representing monthly samples from an annual process.

progress of seasonal processes is often related to degree days, which are calculated in multiple ways and are used, for example, to time agricultural interventions such as the use of pesticides.[5] In such cases, the clock and calendar time of a given experiment can be reconstructed by other researchers only if the starting date and ambient temperatures are reported. Although these conventions have doubtlessly practical advantages, the fact that multiple definitions of time can often not be translated between studies presents methodological challenges and may limit comparisons, meta-analysis, or even spatiotemporally defined repetition.[6] In effect, "time" in these biological processes is not unambiguously and objectively captured by a clock or calendar.

The age of an organism may be similarly ambiguous. Conventionally reported in cumulative counts (e.g., of days following events like fertilization or birth), the biological implications of a given age (i.e., aging) differ between species and between individuals within species. The aging of animals, for example, may proceed more quickly under severe environmental or developmental conditions (see fig. 16.1, *panel B* versus *panel C*).[7] Therefore, the chronological age of an organism, measured in time units since birth, is sometimes distinguished from its biological age, measured by biomarkers associated with physiological processes of aging (see also Chapter 10). The ambiguities of clock and calendar time can reduce the success of replication. For example, if a pharmacological trial involves the re-sampling of mice, doing so after a fixed interval of 100 days seems like an accurate protocol, but differences in individual rates of aging may inflate variation in the outcome (see also Chapter 10). Likewise, if populations of wild animals aged at different rates, their population dynamics could differ because of age-dependent fertility.[8] These problems are not easily resolved because clear biomarkers are rarely available and few studies simultaneously report both chronological and biological age.

When dealing with spatial patterns, it is common to assume that the closer two sampling events are located to each other, the more similar they are to one another. As with analyses of time, spatial autocorrelation generally decreases with distance (i.e., spatial lag) and likewise can be quite variable (see fig. 16.1, *panel A* versus *panel C*).[9] For example, samples collected over short distances but over steep elevational gradients can be very different from one another. Conversely, samples from similar elevations at distant locations (e.g., high-elevation alpine habitats) might be more similar than those from surrounding lowlands, indicating that altitude explains more variation than the distance separating samples.[10]

Even if, in this example, elevation is accounted for, samples or populations from neighboring locations may differ depending on exposure, landscape patch structure, or phylogenetic history.[11] Such obvious examples stand for various other, more subtle, factors that influence the trajectories of spatial autocorrelation. Awareness of these difficulties has resulted in a duplication of space in ecological literature by distinguishing between geographic space (e.g., latitude, longitude) and environmental space (e.g., environmental resources).[12] This duplication is not easily harmonized, as is evident from a study in which biologists from different backgrounds were actually observed and later interviewed.[13] When the researchers were asked to describe both geographic and environmental space, their options became mutually exclusive: there was no single location that fit both descriptions. When working in the field, on the smallest spatial scale, they typically sought to relocate the measurement using a particular device (e.g., a global positioning system [GPS] for measuring geographic space or a pitfall trap for measuring environmental space) to improve the accuracy (often considered to be synonymous to realism) or the representativeness (i.e., generalizability) of the location in space of an organism.[14] In effect, more precise GPS machinery revealed—and in that sense increased—this instance of incommensurability rather than solving it.[15]

Patterns of variation in nature often are periodic rather than (or in addition to) being directional. Accordingly, although the similarity of successive replicates is expected to decrease with the passage of time, it will subsequently increase again (see fig. 16.1, *panel D*). Such periodic processes often relate to geophysical cycles, which arise from movements of the Earth and the Moon around the Sun, and are highly precise and predictable.[16] Chronobiology, the systematic study of adaptations to geophysical cycles, has elaborated that annual and daily, as well as lunar and tidal, cycles fundamentally shape the distribution, behavior, physiology, and gene expression of most, if not all, organisms on earth.[17] Geophysical time scales act in combination with one another. For example, there can be a high probability of encountering an organism in a given state or at a given location depending on a particular combination of phase of solar year, phase of lunar cycle, and time of day.[18] In addition, periodic processes also occur on time scales that are shorter or longer than geophysical cycles and are hence harder to predict. Examples include very short (ultradian) rhythms such as activity bouts of rodents, and longer-term rhythms, including population cycles and the North Atlantic Oscillation (NAO).[19] Repeatability of observations or experiments can be improved if the period and phase of variation in the distribution or state of an organism are taken into account.

Such accounting, however, is also not always straightforward. For example, although dependent on the same geophysical cycle, tidal cycles differ in phase between coasts.[20] Even individuals within species can differ in the phase of their periodic activities (i.e., they have individual chronotypes) or change phase over time, both of which increase variation among replicates and among repeated samples.[21] Additional ambiguity arises from effects of latitude and from precession of geophysical cycles, such as differences in the times of sunrise and sunset across the solar year and among locations.

Organisms often change their activity patterns accordingly, adding complexity to spatial or temporal replicates. For example, sampling of birds during the dawn chorus at a set time in different locations may yield completely different population estimates or assessments of physiology.[22] Researchers might need to decide whether replicates should be carried out according to clock time, light intensity, or the birds' local behavior.

Similar ambiguity applies to periodicities that are not directly linked to geophysical cycles. Population cycles, like those of muskrat and mink, typically differ in phase between locations and often show highly variable periodicities, making suitable replication of a population study very difficult to define.[23]

Anthropogenic changes to natural environments are adding further challenges in accounting for periodic processes. For example, the fast spread of urban sprawl is changing the light environment, and thereby the implication of "night," for increasing numbers of wild organisms.[24] Similar reasoning applies to annual cycles. For example, annual reproduction can differ in timing between years, depending on weather. At present, global climatic change is shifting the timing and progress of annual events (e.g., phenology) in ways that vary across space.[25]

Climatic change affects different organisms differently, so that on a given date in any particular year (e.g., 1 May) components of both the abiotic and the biotic environment may be in different phases relative to earlier studies and to one another. For example, birds that return from migration on the same day may in some years miss the vegetation flush or peak availability of food.[26] Therefore, ecologists have suggested that it may be more informative to relate phenological events, such as timing of reproduction, to variables such as abundance of food instead of solely to clock and calendar time.[27] When data are reported relative to biotic components of time, ecological "distance" (i.e., the "lag" in fig. 16.1) might replace descriptors derived from clocks or calendars. Clearly, without thoughtful choice, comprehensive recording, and translational efforts, the ambiguities pointed out earlier can result in loss of precision and poor success of replication.

COMPONENTS OF TIME AND SPACE: THE IMPORTANCE OF CONCEPTUAL PERSPECTIVES

The many aspects that contribute to the spatiotemporal location of an organism or biological processes collectively determine the probability of its occurrence, the number of individuals present, or its physiological state. Importantly, different fields of biology differ in the emphasis placed on the various aspects in space and time that are studied, in the data collected, and in the ways of quantifying and reporting the data. These differences arise from different conceptual approaches to the study of organisms that are inherent to different research fields and from different criteria—or different priorities among the same criteria—for evaluating the quality of the research and/or data (i.e., epistemic values). However, these differences rarely are reflected explicitly. Whereas all aspects potentially matter, a biologist's concept of time and space affects those aspects that are noted and consequently will inform the design of replicated observations and experiments or repeated studies.

We attempt (table 16.1) to give an overview of aspects of an organism's time and space, grouped into descriptors analogous to commonly used terms in ecology and environmental science (abiotic, biotic, interspecific, intraspecific). We associate these descriptors with three conceptual perspectives, each of which has specific value and meaning in different fields within the life sciences. Hence, these descriptors can be

Table 16.1

Components of time and space that collectively affect the distribution and physiological state of organisms, viewed from three general perspectives: A. exogenous, B. endogenous, and C. interactionist

	TIME	SPACE
A. Exogenous perspective		
Abiotic components		
Unique identifier	Continuous counts from a conventionally set starting point (e.g., religious traditions, software conventions)	Conventional coordinates of latitude and longitude, prime meridian in Greenwich, UK
Underlying geophysical properties	Movements of Earth and Moon, Earth's offset axis: solar annual, diel (daily), lunar, and tidal cycle	Rotational axis of Earth, properties of sphere; location on sphere relative to Earth's axis
Other abiotic components	Environmental seasonality, e.g., long-term temporal patterns of climate, temperature, snow cover, precipitation, etc.; phase of fluctuation in atmospheric pressure; integrals of abiotic factors, e.g., degree days	Elevation/depth, slope, exposure; magnetic field; long-term climate, wind and sea currents
Shared abiotic components	Weather: current local temperature, snow cover, precipitation, humidity, etc.	
Biotic components		
Interspecific biotic components	Timing of other species, e.g., NDVI (normalized difference vegetation index), foliage, progress of agricultural practice, state of succession, predator activity, parasite cycles, food availability	Habitat, species composition, land use, predator density, pathogen load, food availability
Intraspecific biotic components	Activity and state of conspecifics (e.g., flock formation, reproductive state), phase within population cycles	Population density, location relative to lineage, location relative to center of population distribution

(continued)

Table 16.1 (Continued)

	TIME	SPACE
B. Endogenous perspective		
Abiotic components		
Unique identifier	Age (time elapsed, e.g., from birth) Phase (biological clock time of a defined central pacemaker in the brain)	Home (place of origin); destination (e.g., of migration)
Internal properties that determine implications of abiotic components	Endogenous biological timekeeping, with periodicities of annual, diel, lunar, and tidal cycle	Memory of home roost, hibernaculum, wintering site, foraging patch, etc., relative to current position; arguably, internal map
Abiotic properties used as reference for the internal perception	Photic, thermal, gravitational, and other cycles or cues	Photic, geo-magnetic, chemical-gradient, and other cues
Other abiotic components	Temperature, snow cover, precipitation relative to energy supplies, time of resting, available shelter, etc.	
Biotic components		
Internal determinants of implications of interspecific components	Perceived predation risk (e.g., fear), experience of site and time of flowering (time and place learning), susceptibility to parasite (e.g., locally acquired antibodies; seasonal immuno-suppression)	
Internal determinants of implications of intraspecific components	Developmental stage, life-cycle stage, energy reserves, perceived progress of migration; hormonal consequences of social interactions affecting presence and physiology	

C. Interactionist perspective

Abiotic components

Unique identifier	None
Geophysical components resulting from organism's activities	Photic environment as chosen by an organism (e.g., through migration or avoidance of artificial light)
Other abiotic components	Temperature, snow level and precipitation of an organism's constructed microhabitat (e.g., depth and micro-climate of a burrow or nest); tidal height selected by timing of foraging trip

Biotic components

Interspecific biotic components	Pathogen load as a consequence of behavior, food availability inside constructed microhabitat; temporal niche partitioning relative to predators or competitors
Intraspecific biotic components	Self-selected population density; mating status; location, timing, and size of foraging group, experience based on learning, prospecting, eavesdropping, and information sharing

	None
	Latitude and longitude as chosen by an organism (site of coral or of migratory stop-over)

Note: The entries for time and space are intended as illustrative examples, not as an all-inclusive list. Note that several components relate, to different degrees, to both time and space, and are therefore sometimes in a combined entry. For each perspective, entries are ordered from abstract and geophysical to increasingly organism-centered components.

associated with different notations and descriptions (see table 16.1, A–C). We call these perspectives exogenous, endogenous, and interactive. For each perspective, the table is arranged to progress from abstract and geophysical to increasingly organism-centered descriptors.

Exogenous perspective. In the first sense (see table 16.1, A), the location in space and point in time of a manipulation or survey is a framework outside of (exogenous to) the organisms and populations of scientific interest. At one extreme are abstract notions of time and space that are based purely on conventions, such as geographic coordinates based on a system of grid lines conventionally located with respect to the Earth's poles, the Equator, and Greenwich, England, as prime meridian, and time counted from an arbitrary starting point (e.g., the date coded as the number of full days since midnight of 1 January 1900 plus the number of hours, minutes, and seconds for the current day in Microsoft Excel or since January 1904 in Macintosh).

Although there may be good biological reasons for the occurrence of organisms at a given point in time and location in space, we tend to place them in this framework regardless of their specific interests, behavior, or metabolism. The time points and spatial locations denoted by these unique identifiers can then be detailed by further exogenous, abiotic quantifiers. These include latitude and longitude, used separately to characterize gradients; elevations above or below sea level, measured on an arbitrary date; year, day, hour, minute, and second, conventionally decided with respect to Earth's astronomical position relative to the sun; or temperature and other weather factors, measured at times and locations that to some degree relate to the observational or experimental situation.

Some biotic components of time and space also are exogenous in the sense that they are external to the organism, and they capture the context that may be relevant to a specific organism or population, for example, the composition of vegetation or its phenological state. For an organism's time, biotic information, such as the state of vegetation, may be highly relevant, but it rarely is quantified directly. As a shortcut researchers often use indices from remote sensing, such as the normalized difference vegetation index (NDVI).[28] Such indices have to be interpreted with caution: for example, the "greenness" that may be relevant for an ungulate or insectivorous bird may not be apparent from NDVI values, which in turn might reflect the greenness of evergreen tree cover. Likewise, specific biotic features, such as the phenology of a specific host plant or food item, may be a centrally important temporal aspect, whether or not researchers are aware of it. Intraspecifically, a wide range of factors can affect spatiotemporal patterns of distribution and state, for example, availability of a flock or a mate.[29]

Commonly, in spatial analyses, abiotic and biotic information is added in subsequent analytic procedures using well-defined layers in a geographical information system (GIS).[30] As described earlier, ecologists are well aware that there is a crucial ecological dimension to an organism's presence in a geographical location and that, in view of geographical and ecological heterogeneity, proximity in latitude and longitude does not ensure similarity.[31] Hence it may seem to follow that geographic coordinates together with GIS layers can define a location that will suffice for repeatability.

However, this implicit assumption adds uncertainty of its own. Even before adding ecological layers, the geographical coordinates themselves go through a complex georeferencing process, and to estimate its uncertainty one must also record, in the field with a GPS device, the extent, accuracy, and datum of the instrument.[32] Estimates of elevation are also uncertain, as they may be determined either by local barometric pressure or by internal topographic maps. Finally, adding GIS layers of habitat or soil type to more accurately describe a location is far from being globally standardized, and their uncertainty is often not comparable among maps used in different studies. These gaps and uncertainties can be taken into account in statistical models of occupancy and detectability (see Chapter 15); such models are improved with additional specification of how time and space were described throughout the research project and subsequent data management (see Chapters 13, 14). The addition of written narratives and standardized photographs of study sites and times (see Chapters 4, 5) also can be helpful while being open to individual interpretation.

Endogenous perspective. At the other extreme is a concept that is focused on the internal representation of time and space by an individual organism (see table 16.1, B). The presence of an organism, or the state it is in (e.g., reproducing, migrating), may be better understood and described relative to the home range or destination of the organism, to its age, or to a phase in its daily or annual cycle. Likewise, in an experiment, physiological measurements such as the immune response or performance of a study organism may differ depending on an individual's age and circadian clock phase, even if exogenous conditions are identical (see also Chapters 9, 10).[33]

The basis for this organism-centered perspective is an awareness of the importance of physiological and genetic disposition, on the one hand, and of internal representations of time and space, on the other. As explained earlier, many biological processes are scaled to geophysical properties that thereby influence the probability of encountering an organism or of finding it in a specific physiological condition. Because these properties are highly predictable, organisms have evolved internal representations of time and space that direct their behavior and prepare them for upcoming conditions.[34] These internal representations exist separately from external conditions but use the latter as references or cues, so that temporal and spatial behavioral or physiological processes are a combination of internal and external factors.

For example, biological rhythms persist even in completely constant environments, with periodicities close to, but slightly different from, those of corresponding geophysical cycles. Even under captive, constant conditions, animals may orient themselves spatially toward seasonally appropriate destinations or home areas.[35] Internal rhythms may not align fully with environmental predictors. For example, a hibernating species may be active before environmental predictors suggest that it should be, because, based on its internal clock, it may have already emerged from its hibernaculum.[36] Likewise, the presence of a migratory bird in the Sahara Desert may be inferred from knowledge of its migration route and energetic state but not from local exogenous features.[37]

Differences between exogenous and endogenous perspectives can be found for many identified descriptors of time and space (see table 16.1A and B). For example, even occasional predator attacks may change population dynamics and patterns of use of time and space because of perceived predation risk.[38] Ignoring internal mechanisms such as cognitive maps, spatiotemporal effects of experience, memory, and habituation in analysis and modeling have been linked to deviations from predictions and are known to complicate interpretation and replication.[39] As in the case of exogenous factors, considering and documenting endogenous variables can improve the replication of research.

Interactionist perspective. The third intersecting perspective of time and space, which we term interactionist, takes account of the configuration of aspects of time and space by the organisms themselves (see table 16.1, C). This organism-centered approach views the spatiotemporal environment as being dynamically constructed by the actions of organisms over time. Their physiology, metabolism, and in particular behaviors, such as habitat choice, nest building, singing, or feeding, partly define their space and time.[40] In this sense, space and time also become the product of the interaction of the organisms with their environments.

Organisms modulate or configure their spatiotemporal environment through physical interference or movement, among many other activities. For example, the nests and burrows that many organisms construct often lead to temperature and light conditions that differ substantially from those recorded outside and therefore from information derived from GIS layers. Likewise, the choice of a nesting site can affect the reproductive condition as well as the daily and annual timing of a bird.[41] A migratory bird's choice of a wintering area, because of day length or nutritional conditions it encounters, can alter its timing, state, and breeding performance in the ensuing spring in ways not anticipated from local conditions.[42] Cultivation of fungi or storage of food may alter nutrient availability, and consequently population dynamics, but these factors and their effects on replication are not easily captured in ecological surveys and are frequently ignored by species distribution models.[43]

TOWARD RESOLVING AMBIGUITIES OF TIME AND SPACE

Biologists and biological disciplines differ in their concepts of time and space. Community ecologists and biodiversity conservationists normally emphasize exogenous factors; population ecologists and behavioral biologists sometimes focus on interactionist factors; and physiologists may pay closest attention to endogenous factors. All scientists seek to generate results that are generalizable and will be applicable to a wide range of questions, species, times, and places. Yet the difference between concepts of time and space, along with their corresponding research perspectives, can introduce ambiguity and reduce the likelihood of accurately repeating an experiment or observation.

Ignoring the different meanings that denote time and space in one's study is likely to result in empirical gaps and mistakes, whereas pro-actively minding the gaps within each conceptual domain can avoid much of the ambiguity. Resolving the tension requires a more inclusive view that will accommodate more divergent data fields of space (or location) and time on the gradients between the exogenous, endogenous, and interactionist perspectives. We recommend that, where possible, researchers use more than one data model for measuring time and space while keeping track of the variance between the findings from the different data and models (see also Chapters 9, 10, 15 and 17).

To be both specific and general, we suggest six steps one can follow in order to improve the repeatability of observations or experiments. First, be aware of spatiotemporal complexity and diversity. Second, inclusively record such information across its conceptual gradient. Third, if possible, increase sample size and measurements of independent replicates to improve estimates of variability and connectivity of time and space (see Chapters 9, 10, 15). Fourth, record the history of the data and the metadata (see Chapters 13, 14). Fifth, be aware of, and optimize where possible, trade-offs between protocols linked to different concepts of time and space before starting a research project (see examples in Chapters 6–14). Finally, quantify and model the uncertainty as part of assessing data quality (see Chapters 13–15). Although these six steps will not entirely solve the problem of perfect replication over time and space, they provide an excellent start.

NOTES

1. Cressie and Wikle (2011).
2. Shavit and Griesemer (2009, 2011a, 2011b).
3. Schmidt and Starck (2010).
4. Chezik et al. (2013).
5. Rodríguez Caicedo et al. (2012).
6. Chezik et al. (2013).
7. Heidinger et al. (2012); Barrett et al. (2013); Mizutani et al. (2013); Jones et al. (2014).
8. Hammers et al. (2012).
9. Spatial autocorrelation functions normally are displayed as semi-variograms but are directly analogous to temporal autocorrelation plots. See Cressie and Wikle (2011) for additional detail.
10. Körner (2003).

11. Moore et al. (2005); Warren et al. (2014).
12. Gaston et al. (2008).
13. Shavit and Griesemer (2009); Shavit (2016).
14. Here we accept Weisberg's (2006) interpretation of Levins's (1966, 1968) work on modeling strategies in ecology and his claim that Levins uses "accuracy" as if it were synonymous with "realism."
15. Shavit (2016).
16. Foster and Kreitzman (2005).
17. DeCoursey (2004); Foster and Kreitzman (2005); Tessmar-Raible et al. (2011); Numata and Helm (2014).
18. Helm et al. (2012); Zantke et al. (2013).
19. Halle and Stenseth (2000); Haydon et al. (2001); Hueppop and Hueppop (2003); Bloch et al. (2013).
20. Kaiser (2014).
21. Roenneberg et al. (2003).
22. Daan and Aschoff (1975).
23. Haydon et al. (2001).
24. Dominoni et al. (2013a, 2013b).
25. Visser et al. (2010).
26. Visser and Both (2005).
27. Ibid.
28. For examples, see Raess (2008); Tottrup et al. (2012).
29. Filchak et al. (2000); Visser and Both (2005); Helm et.al. (2006); Visser et al. (2010).
30. Store and Jokimäki (2003).
31. Gaston et al. (2008).
32. Wieczorek et al. (2004).
33. Roenneberg et al. (2003); Cermakian et al. (2013).
34. DeCoursey (2004). For examples, see Holland and Helm (2013); Cheeseman et al. (2014).
35. Gwinner and Wiltschko (1980).
36. Lane et al. (2012).
37. Biebach (1985).
38. Zanette et al. (2014).
39. Gremillet and Boulinier (2009); Ruxton and Colegrave (2010).
40. Oyama (1985); Ellison et al. (2005); Shachak et al. (2008).
41. Dominoni et al. (2013a, 2013b).
42. Gwinner and Helm (2003); Studds and Marra (2005).
43. Silva et al. (2003); Guisan and Thuiller (2005); Strickland et al. (2011).

## CHAPTER SEVENTEEN

Best Practices for Creating Replicable Research

Aaron M. Ellison

Research in the biological sciences should be replicable, repeatable, and reproducible. Results should be usable by others and comparable with those from related studies. Data should be traceable from their origin in the field or at the laboratory bench to their deposition in publicly accessible and permanent archives. Technical solutions and best practices already are available to create such open science, but socio-cultural barriers continue to impede its full expression.

Each of the chapters in this book has illustrated a range of conceptual, intellectual, and practical challenges associated with creating replicable, repeatable, and reproducible research. For geographers (Chapter 3), physiologists, ecologists, or environmental scientists working outside of the laboratory (Chapters 4–8, 15), neither replicates within field experiments nor repeated observations or resurveys are likely to yield identical responses to manipulations or patterns of variation. Although laboratories are perceived as scrupulously controlled environments, laboratory studies fare little better than field studies in replicability. Different laboratories working on identical questions regularly have trouble reproducing one another's results: the same laboratory may obtain different results running the same experiment on successive days of the week, and even

genetic identity does not guarantee identical within-experiment replicates or among-experiment reproducibility (Chapters 10–12). It seems
that exact replication may be within the purview only of fictional, obses-
sive personalities such as Borges's Funes, but even his replication turns
out to be a phantasm. For Funes, each object and each process is totally
unique and impossible to replicate, making inference and generalization impossible (Chapter 2).

The seemingly insurmountable challenges of doing precisely
replicable research should not, however, deter scientists either from
repeating experiments or from ensuring that data and results from
observations and experiments are as reliable as possible.[1] Repetition of
well-designed and well-replicated observations, surveys, or experiments
should result in confirmation of robust results (at least within reasonable statistical bounds or limits of precision). Alternatively, the continuing inability to repeat a study may, in the extreme, point toward scientific
misconduct or outright fraud. Full transparency, including detailed description of methods, public availability of data, and thorough documentation of analysis, not only will help researchers interested in repeating a
study and developing general theories but also will reduce incentives to
falsify data and results. The case studies and syntheses that we have presented in this volume, we think, represent steps in the right direction, but
they, too, are imperfect. Recognition of that imperfection is a key part of
improving the practice of doing reproducible research.

Reliable data and the results derived from them also are prerequisite to their use in subsequent scientific research and in providing robust guidance for policies and decisions about pressing issues ranging
from stanching spreading epidemics to ameliorating effects of climatic
change.[2] The confidence and trust that scientists and non-scientists
alike place in scientific findings stem not only from the skeptical stance
that scientists adopt in constantly subjecting research findings—their
own and those of others—to rigorous testing (Chapters 9–10) but also
from an oft-unspoken confidence that the published observations and
experiments had sufficient replication to yield unambiguous results;
the methods used were described in sufficient detail to allow for unambiguous repetition of the study; and the production of final data sets
and their presentation or statistical analysis were transparent and reproducible.

There is no one-size-fits-all set of best practices for creating replicable research in the biological sciences. Taken together, however, the conceptual overviews and case studies presented in this book suggest a suite of reasonable guidelines for creating research that is trustworthy and usable by subsequent researchers. By focusing on the overarching goal of open science that data and results should be transparently documented, usable by others, comparable across studies, and traceable to their source, researchers will create a legacy of a durable, generalizable body of scientific findings and theories.[3]

MINIMAL CONDITIONS FOR REPLICATION

There are four necessary elements that make up the best practices for designing an unbiased, reliable, and trustworthy study (table 17.1). First, the observational or experimental design should account for variation within and among sites, whether in the field or in the laboratory. Second, the acceptable levels of Type I and Type II statistical errors should be set before the project begins. Third, investigators should carefully and quantitatively review the literature and/or do a pilot study to determine the expected effect size (e.g., the expected relationship(s) between two or more variables of interest or the difference in outcome between experimental treatments and controls) and its variance.[4] Finally, there should be a clear assessment of resources available: is there enough time, money, and labor to sample or manipulate the number of replicate samples that are needed to minimize the probability of committing a Type I error and maximize statistical power, given the expected effect size and its variance?

There are no hard and fast rules for the number of individual replicates needed to yield reliable inferences about the population or system of interest.[5] The variance of a sample can be computed from as few as two replicates (Chapters 9–10), but many more are needed to simultaneously minimize the probability of both Type I and Type II statistical errors and, by extension, to maximize statistical power. How many depends directly on the size of the effect of interest, the amount of variability within the population, the acceptable magnitude of the Type I error a researcher is willing to accept (it could be larger than the conventional 5 percent cut-off, the altar of $P \leq 0.05$ at which scientists genuflect

Table 17.1

Minimum conditions for replication within a study and whether each of the case studies presented in previous chapters meets these conditions

CHAPTER AND SUBJECT	ROBUST EXPERIMENTAL DESIGN THAT ACCOUNTS FOR VARIABILITY WITHIN AND AMONG LABORATORIES OR SITES	A PRIORI DEFINITION OF ACCEPTABLE TYPE I AND TYPE II STATISTICAL ERRORS	LITERATURE REVIEW	ASSESSMENT OF AVAILABLE RESOURCES
3. Mapping the human environment	✓	NA	NA	✓
5. Ecological baselines and museum collections	✓	Type I only	✓	✓
7. Terrestrial monitoring	In part	?	✓	✓
8. Marine monitoring	✓	Type I only	✓	✓
10. Immunological experiments with animals and humans	✓	Type I only	✓	✓
12. Clinical trials on the efficacy of antibiotics for treating cholera	✓	✓	✓	✓
14. Bioinformatics: Data provenance and RDataTracker	NA	NA	NA	NA
15. Occupancy modeling	✓	Type I only	✓	✓
16. Quantifying space and time	✓	Type I only	✓	NA

Note: ✓ = meets the condition; ? = not explicitly discussed; NA = not applicable.

by convention), and how much (or how little) statistical power the researcher is willing to accept.[6]

Scientists either accept conventional levels of acceptable Type I and Type II errors (and, by extension, statistical power) or define them before embarking on a research project. Neither is especially difficult to specify a priori; conventional levels of the two types of statistical error are set routinely at 0.05 and 0.2 (power = 0.8), respectively.[7] However, researchers routinely report only their P value (i.e., the probability of committing a Type I error), and most reviewers and journal editors do not expect to see or publish values for statistical power despite widely available software for doing the necessary calculations.[8] Failure to report statistical power does not itself result in research that cannot be repeated successfully, but it does suggest that researchers often are unaware either of the non-linear relationship between Type I error and power or its dependence on the number of observational or experimental replicates.[9]

Researchers also often think that it is difficult to determine effect size and its variance—the other two quantities needed to determine the number of replicates—before doing the study itself. In many cases, however, estimates of effect sizes and their variance can be identified from previously published research or meta-analyses.[10] Alternatively, pilot studies using smaller numbers of replicates can be done first to estimate effect sizes and variance. Either way, the results can be used in a formal power analysis to determine the appropriate number of observational or experimental replicates needed to yield reliable results.[11] If a literature review, pilot study, or a priori power analysis cannot be done, it nonetheless is still important to report the power of the study that was calculated after the research was completed.

Finally, the available resources must be adequate to implement the proposed study design. These should be assessed before one heads into the laboratory or the field, and a decision should be made as to whether the study is even feasible.[12] In short, if a study is worth doing, it is worth doing well.

REPEATABILITY

A study can be designed optimally, have more than adequate replication to minimize Type I statistical errors and maximize statistical power, yield convincing results, and still not be repeatable. Repeating a study means doing it again: making the same measurements; doing the same survey; and running the same experiment at a different time, at either the same place or a different one (e.g., another laboratory, a comparable field site). Variability in conditions, from temperature, atmospheric pressure, or relative humidity among laboratories to differences in vegetation cover or urban development across field sites, ensures that the repeated measurements will not return identical results but only, at best, statistically indistinguishable ones.

The best practice that will enable a successful resurvey or repeated experiment includes the thorough documentation of study methods, including the smallest details (table 17.2). Time and place, in frames of reference appropriate to the organisms or systems under study (Chapter 16), set the initial bounds on repeatability. Until someone invents a time machine, absolute time cannot be repeated, but relative time (e.g., elapsed growing degree days) may be recreated. Geographic locations may be easier to replicate, but physical changes at those locations easily can render a particular place different from what it once was.

Detailed explanation of lab or field methods should include ingredients (e.g., type and magnification of binoculars; steel or aluminum traps used to trap small mammals; cedar or pine shavings for mouse cages; manufacturers of reagents and vaccines) and be written like recipes.[13] Even commercially produced reagents and equipment should be tested independently against standards (Chapter 8).[14] Precision of instruments changes through time as well. Handheld global positioning system receivers have largely replaced maps and compasses for field orientation, but the precision of these receivers may vary from a few centimeters to tens or hundreds of meters. Limitations of space in printed journals that once precluded thorough documentation of methods largely have given way to online supplements or expanded methods sections in scientific papers that are published only online in their entirety. Neither of these sets limits on the lengths of methodological details.

Finally, samples and specimens should be preserved for future analysis. Systematists, taxonomists, evolutionary biologists, and ecologists use physical specimens held in museum collections to determine if they are working on "the same" species (Chapters 4–5). Biomedical researchers similarly depend on inbred lines of plants, animals, or cells

Table 17.2

Criteria for repeatability of observational or experimental studies and whether each of the case studies presented in previous chapters meets these conditions.

CHAPTER AND SUBJECT	DETAILED DESCRIPTION OF METHODS[a]	TIME AND SPACE DEFINED WITH RESPECT TO THE SYSTEM	SAMPLE/SPECIMEN AND ITS DESCRIPTION PRESERVED FOR FUTURE ANALYSIS
3. Mapping the human environment	✓	✓	✓
5. Ecological baselines and museum collections	✓	✓	✓
7. Terrestrial monitoring	✓	No	?
8. Marine monitoring	✓	No	✓
10. Immunological experiments with animals and humans	✓	✓	✓
12. Clinical trials on the efficacy of antibiotics for treating cholera	✓	?	?
14. Bioinformatics: Data provenance and RDataTracker	✓	NA	NA
15. Occupancy modeling	✓	✓	✓
16. Quantifying space and time	✓	✓	✓

Note: ✓ = meets the condition; ? = not explicitly discussed; NA = not applicable.

[a] Includes identification of the person or people who actually did the sampling or collecting.

grown in reliable facilities or on tissue samples and amplified DNA or RNA stored in a range of specimen banks. Maintaining records (specimen labels, pedigrees, sequence data, and the like); quality assurance; and quality control may not be glamorous activities but are essential best practices to ensure the possibility that a study can be repeated.

REPRODUCIBILITY

"Reproducibility" refers to the requirement that data sets can be re-created from the raw data collected by individuals working at the lab bench or in the field or by electronic sensors deployed in the environment.[15] There are at least four technical requirements for creating reproducible data and analyses (table 17.3; Chapters 13–14). First, the data themselves must be publicly available in non-proprietary formats with accurate and precise descriptive metadata. Second, data provenance—process(es) by which usable data sets were generated or derived from raw, often streaming or machine-readable-only data—must be accurately and precisely specified. Third, the computer code ("scripts") and software with which data sets were analyzed must not only be adequately described to ensure their repeated use but also be publicly available in non-proprietary formats. Fourth, version control should be used to ensure that the original data and code are maintained.[16]

These technical requirements can be met relatively easily with current and emerging software and hardware tools.[17] Data should not be stored as spreadsheets created by commercial software and in formats that can be read only by computers equipped with that particular software. Rather, data should be stored as "plain text" with spaces, tabs, commas, or other unique characters separating individual values.[18] Plain-text data files should be accompanied by descriptive metadata that, among other things, accurately and precisely define the format of the data file, the types of variables, the meanings of variable names, and the units of measurement.[19] The descriptive metadata themselves should be machine-readable and should be permanently linked to the data file they describe. Both should be stored in publicly accessible data archives.[20]

Generating, storing, and visualizing process-level metadata (data provenance), while straightforward to describe, is technologically challenging and may require quantities of storage or computational resources

Table 17.3

Reproducibility of data analysis and whether each of the case studies presented in previous chapters meets these conditions. [a]

CHAPTER AND SUBJECT	PUBLIC AND FREE AVAILABILITY OF RAW DATA	COMPLETE DOCUMENTATION OF THE ANALYSIS PROCESS (PROVENANCE OF METADATA)	ADEQUATELY DESCRIBED AND ACCESSIBLE COMPUTER SCRIPTS	VERSION CONTROL
3. Mapping the human environment	✓	✓	NA	✓
5. Ecological baselines and museum collections	✓	No	No	No
7. Terrestrial monitoring	✓	No	No	No
8. Marine monitoring	✓	No	?	No
10. Immunological experiments with animals and humans	?	No	No	No
12. Clinical trials on the efficacy of antibiotics for treating cholera	✓	No	?	No
14. Bioinformatics: Data provenance and RDataTracker	NA	✓	✓	✓
15. Occupancy modeling	✓	Sometimes	Sometimes	No
16. Quantifying space and time	Sometimes	No	No	No

Note: Because the tools and technology that will enable complete and precise reproducibility are still in their infancy, it remains difficult for any researcher to ensure the reproducibility of his or his analysis. But that doesn't mean we shouldn't be trying! ✓ =meets the condition; ? =not explicitly discussed; NA =not applicable.

that far exceed the original analysis (Chapter 14). The seemingly constant upgrading of hardware, operating systems, and software has led to the suggestion that absolute reproducibility of data sets is impossible, regardless of how well data sets and code are documented and archived, but new developments in workflow technology and virtual machines illustrate that creating executable "snapshots" of data sets and accompanying provenance is possible.[21] Such virtual machines are designed to avoid the common problems of software dependencies (i.e., one package relying on many other packages), imprecise documentation (i.e., lack of clear commenting and other metadata), and "code rot" (i.e., the unintended consequence that fixing bugs, upgrading software, and adding new features leads to the inability of prior versions of software or code to be successfully executed).

The primary challenge of implementing requirements for creating open science is cultural: many scientists remain reluctant to share methods or data because of a sense of ownership of an otherwise public (or publicly funded) resource; due to a (usually misplaced) fear that their data and results will be "scooped" if they release data and code prematurely; or because of financial interests and patent protection concerns. Public funding organizations and government agencies in many countries either have no policies on data archiving and availability or actively discourage researchers from making research data available.[22] Secondary challenges include a lack of community-wide standards for what constitutes sufficient methodological detail in published papers or for annotating the diversity of data types generated by biologists; insufficient funding to support the preparation, quality assurance/quality control testing, and accession of data, descriptive metadata, and provenance metadata into permanent, publicly accessible data repositories; and an absence of sanctions—either shared as community norms or enforced by funding agencies—to encourage fully open science and compliance with voluntary or mandated (meta)data archiving. Addressing these challenges has been an active part of the scientific endeavor for hundreds of years. Ongoing interdisciplinary research and action by teams of field scientists and social scientists are helping to align the desired goal of reproducible research with the reward system for career advancement.[23] Future progress depends on active participation by every scientist.

NOTES

Comments from Ayelet Shavit and Julia Jones were of great help in improving this chapter.

1. New initiatives are emerging that propose formal structures to repeat existing studies with the express goal of testing their reproducibility (e.g., Errington et al. 2014).
2. Although the use of scientific evidence in framing policies for these and other issues in the public interest is generally supported, what often remains at issue is the type of evidence that is considered reliable: observational, experimental, or derived from models. See, for example, Rosenberg et al. (2015).
3. It should be appreciated that the creation of open science dates at least to the founding and publication of the first scientific journal, the *Philosophical Transactions of the Royal Society,* in 1665. For additional discussion, see the Open Science entry at *Wikipedia* (https://en.wikipedia.org/wiki/Open _science).
4. Far too often, the litany of citations in the introduction to a research paper is there not because the citations actually have been reviewed for relevance but because the author wants to make sure a potential reviewer or senior colleague is not insulted by seemingly being overlooked in the literature cited.
5. The term "population" does not mean only a single group of organisms. Rather, "population" refers to the larger universe of items that a researcher is interested in studying and from which she draws a sample to observe or use in an experiment. Thus, a "population" not only could be all the red-tailed hawks in Worcester, Massachusetts, but also could be all the small mammals (of all co-occurring species) in Nevada. Similarly, manufacturers might consider all the pencils produced by a factory to be the population of interest to someone who wishes to determine the consistency or quality of a production process.
6. Recall from table 9.1 that power = 1 minus the probability of a Type II error. A useful reference for analysis of non-standard, low-replicate designs is Milliken and Johnson (1989).
7. Because the conventional acceptable level of a Type I error equals 0.05, results are considered statistically "significant" when the observed P value is less than 0.05.
8. For example, the pwr library (Champely et al. 2015) of the R software system, which is based on calculations in Cohen (1988).
9. See Ellison (1996) for discussion and Gotelli and Ellison (2012: Chapter 4, fig. 4.5, for a worked example). Ironically, the well-intentioned efforts to reduce the use of individual animals in biomedical research are thought to have led to the proliferation of a large number of non-repeatable studies. As a result, the major funders of biomedical research in the United Kingdom have revised their guidelines for grant applications that propose using animals (originals in Kilkenny et al. 2010; revisions available online at http://

www.nc3rs.org.uk/arrive-guidelines). The new guidelines require an assessment of statistical power along with a discussion of the research itself. One consequence of these new guidelines may be that more animals may be needed to yield reliable data (Cressey 2015).

10. Such reviews may also reveal that there already are so many data on the topic that an additional experiment cannot be justified (see Chapters 11–12 for additional discussion). From a Bayesian perspective, if the prior information is more informative than will be any new data that could be collected (i.e., the likelihood is dominated by the prior probability distribution), there may be little justification for investing in yet another repetition of a given study (Ellison 1996).

11. Field scientists may be surprised at how low the statistical power is for many ecological studies. If, for example, a measured predictor variable (e.g., temperature) accounts for 50 percent of the variability in an observed response (e.g., timing of flowering), at least 20 replicates are needed to achieve a statistical power of 80 percent. But many published ecological studies have 10 or fewer replicates. The power of many laboratory studies is similarly low: Button et al. (2013) estimated the median power of published research in neuroscience to be only 21 percent.

12. For example, in May 2015 I met a graduate student in South America who was interested in studying the relationship among tree species, the type of soil on which the trees were growing, different forest management practices, and climatic change on the production of volatile organic compounds within the trees' leaves. After discussing the variables of interest and the potential sampling design and running a power analysis, it became apparent that to test the effects of all four variables and their two-, three-, and four-way interactions on volatile production would require thousands of replicates. Not only did the student have only four months in which to do his fieldwork (including not only collecting his samples but also learning to identify the trees in a tropical region where there were >1,000 species in a 25-ha plot), but also there was no high-pressure liquid chromatograph within 1,000 km of his home institution. Finally, he had only sufficient research funds to pay for chromatographic analyses of approximately 100 samples. Such mismatches between design and resources are encountered all too commonly.

13. It is well known, for example, that the cage environment, including the type of wood used in "sterile bedding" for laboratory mice, is associated with a range of responses crucial to reproducing pre-clinical research results (Toth 2015). Not every paper, however, reports the type of sterile bedding used.

14. For example, undeclared or unidentified variability among batches in commercially produced antibodies has led to a remarkably high percentage of experimental results that could not be repeated (Begley and Ellis 2012; Baker 2015).

15. Such sensors and their environments include, among many others, wearable electronic devices designed to monitor the exercise levels or health status of individuals; automated instruments measuring levels of atmospheric carbon dioxide on remote mountaintops; and telescopes deployed on spacecraft orbiting Earth or traveling through space.

16. Even if all of these best practices are followed, however, perfectly reproducible results still can be incorrect (Leek and Peng 2015).

17. Sandve et al. (2013).

18. Such plain-text or ASCII (American Standard Code for Information Interchange) file types include ".txt" (ASCII text, with fields (columns) separated by spaces or tabs) and ".csv" (ASCII text, with fields separated by commas or semicolons). In the United States and many other countries, fractional values are separated from their corresponding integer values with decimal points (periods), and commas are used as field separators in .csv files. In most European and South American countries, decimal points are entered as commas and semi-colons are used as field separators in .csv files. Many spreadsheet packages (e.g., Excel, OpenOffice) allow users to specify field separators, either within each spreadsheet or when exporting files from proprietary binary formats to machine-readable files in plain-text format. Statistical software packages such as R include functions to read .csv files that use either periods or commas as field separators, as well as functions that allow the user to specify the field separator that is used in a given plain-text file. Either way, it is crucial that the descriptive metadata that accompany a data file include an accurate documentation of field separators so as to ensure the readability of individual files and interoperability among data sets.

19. Standards for descriptive metadata exist or are being developed for most scientific disciplines. Relevant examples for biologists include EML for the ecological sciences (Michener et al. 1997), the Genomic Contextual Data Markup Language (Gil et al. 2008), a simple annotation format for data from the Library of Integrated Network-based Cellular Signatures of the US National Institutes of Health (Vempati et al. 2014), and the Dublin Core for heterogeneous data (http://dublincore.org/), among many others.

20. Some sub-disciplines in the biological sciences have unique archives for particular data types (e.g., GenBank for genetic sequence data: http://www.ncbi.nlm.nih.gov/). Others have a diversity of archives (e.g., ecological data can be accessed into Dryad (http://www.datadryad.org), the Distributed Active Archive Center for Biogeochemical Dynamics (http://daac.ornl.gov), or the Long Term Ecological Research Network's Network Information System (https://portal.lternet.edu/nis/home.jsp). Even data that should be protected or embargoed from general use (e.g., data on locations of endangered species, individual identities in genetic studies) should be accessed into data archives. Such protection should be rarely applied and fully justified.

21. Boettiger (2014).

22. Individual reluctance to make data publicly available or official positions dis-couraging data archiving will effectively marginalize individual scientists and countries. More and more international journals are signing on to guidelines to promote open research and requiring submission of data and analytical code (Nosek et al. 2015). Researchers who do not make their data available simply will be unable to publish in reputable international jour-nals.

23. See, for example, Ioannidis et al. (2014); Miguel et al. (2014).

Epilogue

A CHORUS'S DANCE WITH REPLICATION

Ayelet Shavit

If you can't get rid of the skeleton in your closet, you'd best teach it to dance.

—*George Bernard Shaw*

After reviewing the many different pitfalls on the road to replication, one might conclude that any attempt to truly replicate a scientific study is a tragedy waiting to happen. No matter how hard one tries, due to the different goals of replication—tracking change, accurately describing and analyzing a biological system, producing a generally representative diagnosis or prediction from this particular system—and their complex interactions, it appears that there would always remain a few skeletons rattling in the scientific closet. However, throughout this book we have argued to the contrary, that is, that exact replication may not be a requirement of good science but that planning ahead to recognize, record, and—whenever possible—quantify empirical discontinuities and conceptual ambiguities regarding replication is the key to curing "the plague of non-reproducibility in science."[1] To conclude this book, we provide and suggest the use of a practical flowchart for interpreting the best practices for replication (fig. E.1). Taking the specific actions shown in the flowchart will help researchers to bridge, albeit not completely and permanently close, the gaps inherent in replication.

At each branch point, making the "wrong" decision—for example, ignoring (that is, not recording) or conflating (that is, not recording separately) the relevant details—closes the door to replication. Making the "right" decision, however, at best only clarifies and quantifies how much further away we remain from exact replication. Either way, the hubris implicit in any attempt to perfectly replicate a project is fated to fail. Furthermore, philosophical (Chapter 2), historical (Chapter 3), and theoretical (Chapter 15) discussions of replication suggest that full replication is not even a desirable goal in the life sciences. In that sense, the flowchart echoes a song of a Greek chorus, summarizing the different voices heard throughout this book and structured as a logical decision tree.

It is no accident that this book ends with tears. In ancient Greek theater, tragedy could befall only *the hero:* the man or woman of virtue who truly and rigorously attempted to do everything right but with each small choice came closer and closer to inevitable failure. Uncovering the hero's hidden flaw re-established social order in the world. The resulting catharsis provided emotional relief to the audience while leaving intact rational doubts about the social order itself.[2] The Greek chorus played a central role in presenting classical tragedies. A group of performers, embodying a collective character, were responsible for building a common place and time in which the story unfolded, narrating the thoughts and feelings of the other actors, summarizing moral lessons for the audience, and pausing the drama at crucial points to allow the members of the audience to critically reflect on these lessons.[3] Reflections by scientists on how best to solve, or work around, the problem of replication have been echoed by the common threads of this conundrum illuminated by philosophers and historians.

The flowchart ends by identifying the time to end a repeated experiment or analysis. In closing this book, we hope that the assembly of different disciplinary voices has produced a musical theme worth listening to.

Yet listening is not enough, at least not in a tragedy. In Greek plays, the chorus was located at the orchestra, which literally means "the dancing place." Throughout the tragedy, the chorus not only sang but also, and perhaps mostly, danced. Through his extensive knowledge of Greek culture, Nietzsche tempted us to dance at the edge of our abyss, writing: "We should consider every day lost on which we have not danced at least once. And we should call every truth false which was not accompanied by at least one laugh."[4]

Our play has ended; the observations and experiments have been performed, the results published, and the audience has gone home. We

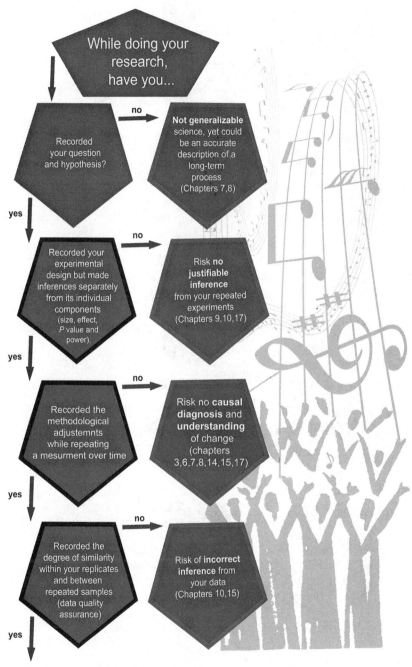

Fig. E.1. Flowchart for identifying and tackling problems of replication.
Figure created by Tamar Ben Barush, and used with permission.

(continued)

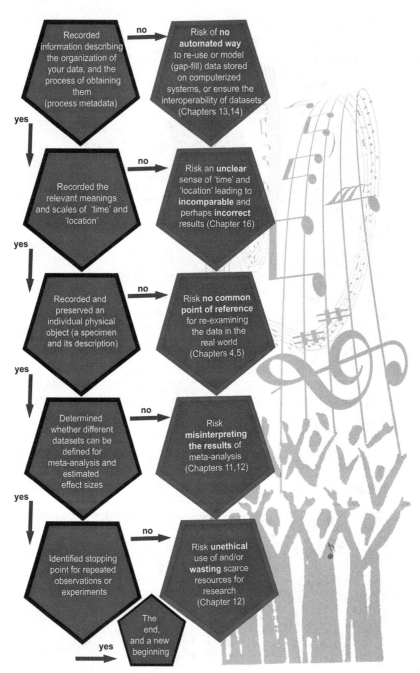

Fig. E.1 (continued)

hope that you are gently laughing as you turn this page, ready to commence your next scientific dance.

NOTES

This chapter has improved immensely due to the creative ideas and cooperation of Tommer Rotstein, Bar Katin, Sivan Margalit, Iris Eshed, and Yael Silver.

1. Hayden (2013).
2. Weiner (1980).
3. Ibid.
4. Nietzsche (1977: 260).

Abdi, H. 2007. "Bonferroni and Šidák Corrections for Multiple Comparisons." Pages 103–107 in *Encyclopedia of Measurement and Statistics,* edited by N. J. Salkind. Thousand Oaks, CA: Sage.

Ahumada, J. A., C.E.F. Silva, K. Gajapersad, C. Hallam, et al. 2011. "Community Structure and Diversity of Tropical Forest Mammals: Data from a Global Camera Trap Network." *Philosophical Transactions of the Royal Society B: Biological Sciences,* 366: 2703–2711.

Alberts, B., et al. 2015. "Self-correction in Science at Work: Improve Incentives to Support Research Integrity." *Science* 348: 1420–1422.

Agassi, J., and A. Meidan. 2008. *Philosophy from a Sceptical Perspective.* New York: Cambridge University Press.

Aldrich, E., and J.F.A. Symonds. 1841. *Plan of the Town and Environs of Jerusalem . . . from the Original Drawing of the Survey Made in the Month of March 1841.* London: James Wyld.

Andorno, R. 2004. "The Precautionary Principle: A New Legal Standard for a Technological Age." *Journal of International Biotechnology Law* 1: 11–19.

Andreakis N., G. Procaccini, C. Maggs, and W.H.C.F. Kooistra. 2007. "Phylogeography of the Invasive Seaweed *Asparagopsis* (Bonnemaisoniales, Rhodophyta) Reveals Cryptic Diversity." *Molecular Ecology* 16: 2285–2299.

Angel, D. L., S. Verghese, J. J. Lee, A. M. Saleh, et al. 2000. "Impact of a Net Cage Fish Farm on the Distribution of *Benthic foraminifera* in the Northern Gulf of Eilat (Aqaba, Red Sea)." *Journal of Foraminiferal Research* 30: 54–65.

Bacher, S. 2012. "Still Not Enough Taxonomists: Reply to Joppa et al." *Trends in Ecology and Evolution* 27: 65–66.

Baird, R. C. 2010. "Leveraging the Fullest Potential of Scientific Collections through Digitisation." *Biodiversity Informatics* 7: 130–136.

Baker, M. 2015. "Reproducibility Crisis: Blame It on the Antibodies." *Nature* 521: 274–276.

Barrett, E. L., T. A. Burke, M. Hammers, J. Komdeur, et al. 2013. "Telomere Length and Dynamics Predict Mortality in a Wild Longitudinal Study." *Molecular Ecology* 22: 249–259.

Bartlett, J. R. 2008. "Mapping Jordan through Two Millennia." *Palestine Exploration Fund Annual* 10: 100–103.

Barzilay, E. J., N. Schaad, N. R. Magloire, K. S. Mung, et al. 2013. "Cholera Surveillance during the Haiti Epidemic—The First 2 Years." *New England Journal of Medicine* 368: 599–609.

Bassalla, G. 1967. "The Spread of Western Science." *Science* 156: 611–622.

Baumer, B., M. Cetinkava-Rundel, A. Bray, L. Loi, et al. 2014. "R Markdown: Integrating a Reproducible Analysis Tool into Introductory Statistics." Available at http://arxiv.org/pdf/1402.1894v1. Accessed 29 June 2016.

Bayes, T. 1763. "An Essay towards Solving a Problem in the Doctrine of Chances: By the Late Rev. Mr. Bayes, Communicated by Mr. Price, in a Letter to John Canton, M. A. and F. R. S." *Philosophical Transactions of the Royal Society of London* 53: 370–418.

Beever, E. A. 2006. "Monitoring Biological Diversity: Strategies, Tools, Limitations and Challenges." *Northwestern Naturalist* 87: 66–79.

Begley, C. G., and L. M. Ellis. "Drug Development: Raise Standards for Preclinical Cancer Research." *Nature* 483: 531–533.

Begley, C. G., A. M. Buchan, and U. Dirnagl. 2015. "Robust Research: Institutions Must Do Their Part for Reproducibility." *Nature* 525: 25–27.

Begon, M., C. R. Townsend, and J. L. Harper. 2006. *Ecology: From Individuals to Ecosystems*, 4th edition. Oxford: Blackwell.

Belbin, L., J. Daly, T. Hirsch, D. Hobern, et al. 2013. "A Specialist's Audit of Aggregated Occurrence Records: An 'Aggregator's' Perspective." *ZooKeys* 305: 67–76.

Belnap, J. 1998. "Choosing Indicators of Natural Resource Condition: A Case Study in Arches National Park, Utah, USA." *Environmental Management* 22: 635–642.

Ben-Arieh, Y. 1972. "Pioneer Scientific Exploration in the Holy Land at the Beginning of the Nineteenth Century." *Terrae Incognitae* 4: 95–110.

——. 1973. "The First Surveyed Maps of Jerusalem." *Eretz-Israel: Archaeological, Historical and Geographical Studies* 11: 64–74 [in Hebrew].

——. 1974. "The Catherwood Map of Jerusalem." *Quarterly Journal of the Library of Congress* 31: 150–160.

——. 1979. *The Rediscovery of the Holy Land in the Nineteenth Century*. Jerusalem: Magnes.

——. 1987. "Perceptions and Images of the Land of Israel in the Writings of Nineteenth-Century Western Travelers." Pages 89–114 in *Transition and*

Change in Modern Jewish History: Essays Presented in Honor of Shmuel Ettinger, edited by S. Almog. Jerusalem: Historical Society of Israel [in Hebrew].

Berg, N., and A. Perevolotsky. 2011. *A National Program for Biodiversity Monitoring.* Jerusalem: HAMMARAG, The Israeli Academy of Science [in Hebrew].

Berg, T. P., and G. Meulemans. 1991. "Acute Infectious Bursal Disease in Poultry: Protection Afforded by Maternally Derived Antibodies and Interference with Live Vaccination." *Avian Pathology: Journal of the World Veterinary Poultry Association* 20: 409–421.

Berghaus, H. 1835. *Geographisches Memoir zur Erklärung und Erläuterung der Karte von Syrien.* Gotha, Germany: Justus Perthes.

Benjamini, Y., and Y. Hochberg. 1995. "Controlling the False Discovery Rate: A Practical and Powerful Approach to Multiple Testing." *Journal of the Royal Statistical Society, Series B,* 57: 289–300.

Bhattacharya, S. K., M. K. Bhattacharya, P. Dutta, D. Dutta, et al. 1990. "Double-Blind, Randomized, Controlled Clinical Trial of Norfloxacin for Cholera." *Antimicrobial Agents and Chemotherapy* 34: 939–940.

Biebach, H. 1985. "Sahara Stopover in Migratory Flycatchers: Fat and Food Affect the Time Program." *Experientia* 41: 695–697.

Bisby, F. A. 2000. "The Quiet Revolution: Biodiversity Informatics and the Internet." *Science* 289: 2309–2312.

Bloch, G., B. M. Barnes, M. P. Gerkema, and B. Helm. 2013. "Animal Activity around the Clock with No Overt Circadian Rhythms: Patterns, Mechanisms and Adaptive Value." *Proceedings of the Royal Society B: Biological Sciences,* 280: 20130019.

Boettiger, C. 2014. "An Introduction to Docker for Reproducible Research, with Examples from the R Environment." Available at http://arxiv.org/pdf/1410 .0846v1.pdf. Accessed 29 June 2016.

Bolland, M. J., A. Grey, G. D. Gamble, and I. R. Reid. 2014. "Vitamin D Supplementation and Falls: A Trial Sequential Meta-analysis. *Lancet Diabetes and Endocrinology* 2: 573–580.

Bookman, R., Y. Bartov, Y. Enzel, and M. Stein. 2006. "Quaternary Lake Levels in the Dead Sea Basin: Two Centuries of Research." *Geological Society of America Special Papers* 401: 155–170.

Bookman, R., S. Filin, Y. Avin, D. Rosenfeld, et al. 2014. "Possible Connection between Large Volcanic Eruptions and Level Rise Episodes in the Dead Sea Basin." *Quaternary Science Reviews* 89: 123–128.

Boose, E. R., A. M. Ellison, L. J. Osterweil, R. Podorozhny, et al. 2007. "Ensuring Reliable Datasets for Environmental Models and Forecasts." *Ecological Informatics* 2: 237–247.

Borell, A. E., and R. Ellis. 1934. "Mammals of the Ruby Mountains Region of Northeastern Nevada." *Journal of Mammalogy* 15: 12–44.

Borges, J. L. 1998. *Collected Fictions.* New York: Penguin.

——. 1999. *Selected Non-fictions.* New York: Viking.

Borgs, L., P. Beukelaers, R. Vandenbosch, S. Belachew, et al. 2009. "Cell 'Circadian' Cycle: New Role for Mammalian Core Clock Genes." *Cell Cycle* 8: 832–837.

Bortolus, A. 2008. "Error Cascades in the Biological Sciences: The Unwanted Consequences of Using Bad Taxonomy in Ecology." *Ambio* 37: 114–118.

Bowers, S., T. McPhillips, and B. Ludäscher. 2012. "Declarative Rules for Inferring Fine-Grained Data Provenance from Scientific Workflow Execution Traces." Pages 82–96 in *Provenance and Annotation of Data and Processes,* edited by P. Groth and J. Frew. Berlin: Springer.

Bowker, G. C. 2005. *Memory Practices in the Sciences.* Cambridge, MA: MIT Press.

Box, G.E.P., and N. R. Draper. 1987. *Empirical Model Building and Response Surfaces.* New York: John Wiley and Sons.

Brown, J. H., and B. A. Maurer. 1989. "Macroecology—The Division of Food and Space among Species on Continents." *Science* 243: 1145–1150.

Bruno, J. F., and E. R. Selig. 2007. "Regional Decline of Coral Cover in the Indo-Pacific: Timing, Extent, and Subregional Comparisons." *PLOS ONE* 2: e711.

Brunt, J. W. 2000. "Data Management Principles, Implementation and Administration." Pages 25–47 in *Ecological Data: Design, Management, Processing,* edited by W. K. Michener and J. W. Brunt. Oxford: Blackwell Science.

Buck, S. 2015. "Solving Reproducibility." *Science* 348: 1403.

Buckland, S. T., A. E. Magurran, R. E. Green, and R. M. Fewster. 2005. "Monitoring Change in Biodiversity through Composite Indices." *Philosophical Transactions of the Royal Society B: Biological Sciences,* 360: 243–254.

Bunce, R.G.H., M. J. Metzger, R.H.G. Jongman, J. Brandt, et al. 2008. "A Standardized Procedure for Surveillance and Monitoring European Habitats and Provision of Spatial Data." *Landscape Ecology* 23: 11–25.

Burans, J. P., J. Podgore, M. M. Mansour, A. H. Farah, et al. 1989. "Comparative Trial of Erythromycin and Sulphatrimethoprim in the Treatment of Tetracycline-Resistant *Vibrio cholerae* O1." *Transactions of the Royal Society, Tropical Medicine and Hygiene,* 83: 836–838.

Butler, T., S. Lolekha, C. Rasidi, A. Kadio, et al. 1993. "Treatment of Acute Bacterial Diarrhea: A Multicenter International Trial Comparing Placebo with Fleroxacin Given as a Single Dose or Once Daily for 3 days." *American Journal of Medicine* 94: 187S–194S.

Button, K. S., J. P. Ioannidis, C. Mokrysz, B. A. Nosek, et al. 2013. "Power Failure: Why Small Sample Size Undermines the Reliability of Neuroscience." *Nature Reviews Neuroscience* 14: 365–376.

Cabello, J., N. Fernández, D. Alcaraz-Segura, C. Oyonarte, et al. 2012. "The Ecosystem Functioning Dimension in Conservation: Insights from Remote Sensing." *Biodiversity and Conservation* 21: 3287–3305.

Cahill, J. F. Jr., J. P. Castelli, and B. B. Casper. 2002. "Separate Effects of Human Visitation and Touch on Plant Growth and Herbivory in an Old-Field Community." *American Journal of Botany* 89: 1401–1409.

Campbell, J. L., L. E. Rustad, J. H. Porter, J. R. Taylor, et al. 2013. "Quantity Is Nothing without Quality: Automated QA/QC for Streaming Sensor Networks." *BioScience* 63: 574–585.

Carpenter, C. C., R. B. Sack, A. Mondal, and P. P. Mitra. 1964. "Tetracycline Therapy in Cholera." *Journal of the Indian Medical Association* 143: 309–312.

Carpenter, S. R. 1989. "Replication and Treatment Strength in Whole-Lake Experiments." *Ecology* 70: 453–463.

——. 1990. "Large-Scale Perturbations: Opportunities for Innovation." *Ecology* 71: 2038–2043.

Cartwright, N. 1989. *Nature's Capacities and Their Measurement.* Oxford: Clarendon.

——. 2002. "In Favor of Laws That Are Not *Ceteris Paribus* After All." *Erkenntnis* 75: 425–439.

——. 2012. *Evidence: For Policy and Wheresoever Rigor Is a Must.* London: The London School of Economics and Political Science.

Cassey, P., and T. M. Blackburn. 2006. "Reproducibility and Repeatability in Ecology." *BioScience* 56: 958–959.

Cermakian, N., T. Lange, D. Golombek, D. Sarkar, et al. 2013. "Crosstalk between the Circadian Clock Circuitry and the Immune System." *Chronobiology International* 30: 870–888.

Chamberlin, T. C. 1890. "The Method of Multiple Working Hypotheses." *Science* (Old Series) 15: 92–96.

Champely, S., C. Ekstrom, P. Dalgaard, J. Gill, et al. 2015. "Package 'pwr': Basic Functions for Power Analysis." The Comprehensive R Archive Network. Available at https://cran.r-project.org/web/packages/pwr/pwr.pdf. Accessed 29 June 2016.

Chapman, A. D. 2005. *Principles of Data Quality, Version 1.0.* Copenhagen: Global Biodiversity Information Facility. Available at http://www.gbif.org /orc/?doc_id=1229. Accessed 29 June 2016.

Chapman, A. D., and J. Wieczorek, editors. 2006. "Biogeomancer: Guide to Best Practices in Georeferencing." Copenhagen: Global Biodiversity Information Facility. Available at http://www.gbif.org/orc/?doc_id=1288. Accessed 29 June 2016.

Charron-Bost, B., F. Pedone, and A. Schiper, editors. 2010. *Replication: Theory and Practice* (Lecture Notes from Computer Science 5959). Berlin: Springer.

Cheeseman, J. F., C. D. Millar, U. Greggers, K. Lehmann, et al. 2014. "Way-Finding in Displaced Clock-Shifted Bees Proves Bees Use a Cognitive Map." *Proceedings of the National Academy of Sciences, USA,* 111: 8949–8954.

Chezik, K. A., N. P. Lester, and P. A. Venturelli. 2014. "Fish Growth and Degree-Days II: Selecting a Base Temperature for an Among-Population Study." *Canadian Journal of Fisheries and Aquatic Sciences* 71: 1303–1311.

Cohen, J. 1988. *Statistical Power Analysis for the Behavioral Sciences,* 2nd edition. Hillsdale, NJ: Lawrence Erlbaum.

Collins, H. M. 1992. *Changing Order, Replication and Induction in Scientific Practice*. Chicago: University of Chicago Press.

Conan Doyle, A. 1890. *The Sign of Four*. London: Spencer Blackett.

Contamin, R., and A. M. Ellison. 2009. "Indicators of Regime Shifts in Ecological Systems: What Do We Need to Know and When Do We Need to Know It?" *Ecological Applications* 19: 799–816.

Cook, B. I., T. R. Ault, and J. E. Smerdon. 2015. "Unprecedented 21st Century Drought Risk in the American Southwest and Central Plains." *Science Advances* 1: e1400082.

Cooper-Ellis, S., D. R. Foster, G. Carlton, and A. L. Lezberg. 1999. "Forest Response to Catastrophic Wind: Results from an Experimental Hurricane." *Ecology* 80: 2683–2696.

Costanza, R., R. d'Arge, R. de Groot, S. Farberk, et al. 1997. "The Value of the World's Ecosystem Services and Natural Capital." *Nature* 387: 253–260.

Costello, M. J., and J. Wieczorek. 2014. "Best Practice for Biodiversity Data Management and Publication." *Biological Conservation* 173: 68–73.

Costello, M. J., W. K. Michener, M. Gahegan, Z. Q. Zhang, et al. 2013. "Biodiversity Data Should Be Published, Cited, and Peer Reviewed." *Trends in Ecology and Evolution* 28: 454–461.

Cressey, D. 2015. "UK Funders Demand Strong Statistics for Animal Studies." *Nature* 520: 271–272.

Cressie, N., and C. K. Wikle. 2011. *Statistics for Spatio-temporal Data*. Hoboken, NJ: John Wiley and Sons.

Daan, S., and J. Aschoff. 1975. "Circadian Rhythms of Locomotor Activity in Captive Birds and Mammals: Their Variations with Season and Latitude." *Oecologia* 18: 269–316.

Dale, V. H., and S. C. Beyeler. 2001. "Challenges in the Development and Use of Ecological Indicators." *Ecological Indicators* 1: 3–10.

Dalton, R. 2007. "Endangered Collections." *Nature* 446: 605–606.

Darwin, C. 1859. *On the Origin of Species by Means of Natural Selection*. New York: The Heritage (1963 edition).

Davidson, D. 1967. "Causal Relations." *Journal of Philosophy* 64: 691–703.

Davies, W.K.D. 1968. "The Need for Replication in Human Geography: Some Central Place Examples." *Tijdschrift voor Economische en Sociale Geografie* 59: 145–155.

Dawson, T. P., S. T. Jackson, J. I. House, I. C. Prentice, et al. 2011. "Beyond Predictions: Biodiversity Conservation in a Changing Climate." *Science* 332: 53–58.

de Bertou, J. 1839. "Itinéraire de la Mer Morte à Akaba par les Wadys-el-Ghor, el-Araba et el-Akaba, et retour à Hébron par Petra." *Bulletin de la Société de Géographie de Paris* 11: 274–331.

Debinski, D. M., and R. D. Holt. 2000. "A Survey and Overview of Habitat Fragmentation Experiments." *Conservation Biology* 14: 342–355.

De Boeck, H. J., M. Liberloo, B. Gielen, I. Nijs, et al. 2008. "The Observer Effect in Plant Science." *New Phytologist* 177: 579–583.

de Carvalho, M. R., F. A. Bockmann, D. S. Amorim, and C.R.F. Brandao. 2008. "Systematics Must Embrace Comparative Biology and Evolution, Not Speed and Automation." *Evolutionary Biology* 35: 150–157.

DeCoursey, P. 2004. "The Behavioral Ecology and Evolution of Biological Timing Systems." Pages 27–66 in *Chronobiology: Biological Timekeeping*, edited by J. C. Dunlap, J. J. Loros, and P. DeCoursey. Sunderland, MA: Sinauer Associates.

Desombere, I., T. Cao, Y. Gijbels, and G. Leroux-Roels. 2005. "Non-responsiveness to Hepatitis B Surface Antigen Vaccines Is Not Caused by Defective Antigen Presentation or a Lack of B7 Co-stimulation." *Clinical and Experimental Immunology* 140: 126–137.

Diamond, J. M. 1972. *Avifauna of the Eastern Highlands of New Guinea*. Cambridge, MA: Nuttall Ornithological Club.

———. 1986. "Overview: Laboratory Experiments, Field Experiments, and Natural Experiments." Pages 3–22 in *Community Ecology*, edited by J. Diamond and T. J. Case. New York: Harper and Row.

Diamond, J. M., and M. Lecroy. 1979. "Birds of Karkar and Bagabag Islands, New Guinea." *Bulletin of the American Museum of Natural History* 164: 467–532.

Diebold, F. X. 2012. "A Personal Perspective on the Origin(s) and Development of 'Big Data': The Phenomenon, the Term, and the Discipline." *Penn Institute for Economic Research* 13-003. Philadelphia: University of Pennsylvania. Available at http://www.ssc.upenn.edu/~fdiebold/papers/paper112/Diebold_Big_Data.pdf. Accessed 29 June 2016.

DiNardo, J. 2008. "Natural Experiments and Quasi-Natural Experiments." In *The New Palgrave Dictionary of Economics*, 2nd edition, edited by S. N. Durlauf and L. E. Blume. London: Palgrave Macmillan. Available at http://dx.doi.org/10.1057%2F9780230226203.1162. Accessed 13 March 2014.

Doak D. F., J. A. Estes, B. S. Halpern, U. Jacob, et al. 2008. "Understanding and Predicting Ecological Dynamics: Are Major Surprises Inevitable?" *Ecology* 89: 952–961.

Doak, D. F., V. J. Bakker, B. E. Goldstein, and B. Hale. 2014. "What Is the Future of Conservation?" *Trends in Ecology and Evolution* 29: 77–81.

Doctorow, C. 2001. "Metacrap: Putting the Torch to Seven Straw-men of the Meta-utopia." Self-published. Available at http://www.well.com/~doctorow/metacrap.htm. Accessed 21 January 2015.

Doja, A., and W. Roberts. 2006. "Immunizations and Autism: A Review of the Literature." *Canadian Journal of Neurological Sciences* 33: 341–346.

Dolev, A., and A. Perevolotsky. 2004. *The Red Book: Vertebrates in Israel*. Jerusalem: Israel Nature and Parks Authority and the Society for Protection of Nature in Israel.

Dominoni, D. M., B. Helm, M. Lehmann, H. B. Dowse, et al. 2013a. "Clocks for the City: Circadian Differences between Forest and City Songbirds." *Proceedings of the Royal Society B: Biological Sciences*, 280: 20130593.

Dominoni, D. M., M. Quetting, and J. Partecke. 2013b. "Artificial Light at Night Advances Avian Reproductive Physiology." *Proceedings of the Royal Society B: Biological Sciences,* 280: 20123017.

Drew, J. 2011. "The Role of Natural History Institutions and Bioinformatics in Conservation Biology." *Conservation Biology* 25: 1250–1252.

Duhem, P. 1906. *The Aim and Structure of Physical Theory.* Princeton, NJ: Princeton University Press (1954 edition).

———. 1969. *To Save the Phenomena,* translated by E. Doland and C. Maschler. Chicago: University of Chicago Press.

Duro, D. C., N. C. Coops, M. A. Wulder, and T. Han. 2007. "Development of a Large Area Biodiversity Monitoring System Driven by Remote Sensing." *Progress in Physical Geography* 31: 1–26.

Dutta, D., S. K. Bhattacharya, M. K. Bhattacharya, A. Deb, et al. 1996. "Efficacy of Norfloxacin and Doxycycline for Treatment of *Vibrio cholerae* 0139 Infection." *Journal of Antimicrobial Chemotherapy* 37: 575–581.

Dwan, K., D. G. Altman, J. A. Arnaiz, J. Bloom, et al. 2008. "Systematic Review of the Empirical Evidence of Study Publication Bias and Outcome Reporting Bias." *PLOS ONE* 3: e3081.

Ebach, M. C., A. G. Valdecasas, and Q. D. Wheeler. 2011. "Impediments to Taxonomy and Users of Taxonomy: Accessibility and Impact Evaluation." *Cladistics* 27: 550–557.

Edney, M. H. 1997. *Mapping an Empire: The Geographical Construction of British India, 1765–1843.* Chicago: University of Chicago Press.

Edwards, J. L. 2004. "Research and Societal Benefits of the Global Biodiversity Information Facility." *BioScience* 54: 485–486.

Egea, E., A. Iglesias, M. Salazar, C. Morimoto, et al. 1991. "The Cellular Basis for Lack of Antibody Response to Hepatitis B Vaccine in Humans." *Journal of Experimental Medicine* 173: 531–538.

Elkana, Y. 1974. *The Discovery of the Conservation of Energy.* London: Hutchinson.

Ellis, G., and J. Silk. 2014. "Defend the Integrity of Physics." *Nature* 516: 321–323.

Ellison, A. M. 1996. "An Introduction to Bayesian Inference for Ecological Research and Environmental Decision-making." *Ecological Applications* 6: 1036–1045.

Ellison, A. M., and B. Dennis. 2010. "Paths to Statistical Fluency for Ecologists." *Frontiers in Ecology and the Environment* 8: 362–370.

Ellison, A. M., M. S. Bank, B. D. Clinton, E. A. Colburn, et al. 2005. "Loss of Foundation Species: Consequences for the Structure and Dynamics of a Forest Ecosystem." *Frontiers in Ecology and the Environment* 3: 479–486.

Ellison, A. M., L. J. Osterweil, L. Clarke, J. L. Hadley, A. Wise, et al. 2006. "Analytic Webs Support the Synthesis of Ecological Datasets." *Ecology* 87: 1345–1358.

Ellison, A. M., N. J. Gotelli, B. D. Inouye, and D. R. Strong. 2014. "*P* Values, Hypothesis Testing, and Model Selection: It's Déjà Vu All Over Again." *Ecology* 95: 609–610.

Ellison, A. M., N. J. Gotelli, and P. Siqueira. Under contract. *Big Data; Design and Analysis of Large-Scale Studies in Ecology and Environmental Science.* Sunderland, MA: Sinauer Associates.

Ellkvist, T., D. Koop, C. Silva, J. Freire, et al. 2009. "Using Mediation to Achieve Provenance Interoperability." Pages 291–298 in *Proceedings of IEEE 2009 Third International Workshop on Scientific Workflows* (SWF 2009). Los Angeles: IEEE Computer Society.

Elmendorf, S. C., G.H.R. Henry, R. D. Hollister, A. M. Fosaa, et al. 2015. "Experiment, Monitoring, and Gradient Methods Used to Infer Climate Change Effects on Plant Communities Yield Consistent Patterns." *Proceedings of the National Academy of Sciences, USA,* 112: 448–452.

Elzinga, C. L., D. W. Salzer, J. W. Willoughby, and J. P. Gibbs. 2001. *Monitoring Plant and Animal Populations.* Oxford: Blackwell.

Engemann, K., B. J. Enquist, B. Sandel, B. Boyle, et al. 2015. "Limited Sampling Hampers "Big Data" Estimation of Species Richness in a Tropical Biodiversity Hotspot. *Ecology and Evolution* 5: 807–820.

Errington, T. M., E. Iorns, W. Gunn, F. E. Tan, et al. 2014. "An Open Investigation of the Reproducibility of Cancer Biology Research." *eLife* 3: e04333.

Evers, D. C., J. A. Schmutz, N. Basu, C. R. DeSorbo, et al. 2014. "Historic and Contemporary Mercury Exposure and Potential Risk to Yellow-Billed Loons (*Gavia adamsii*) Breeding in Alaska and Canada." *Waterbirds* 37: 147–159.

Falagas, M. E., I. K. Kotsantis, E. K. Vouloumanou, and P. I. Rafailidis. 2009. "Antibiotics versus Placebo in the Treatment of Women with Uncomplicated Cystitis: A Meta-analysis of Randomized Controlled Trials." *Journal of Infection* 58: 91–102.

Fancy, S. G., J. E. Gross, and S. L. Carte. 2009. "Monitoring the Condition of Natural Resources in US National Parks." *Environmental Monitoring and Assessment* 151: 161–174.

Farnsworth, E. J. 2014. "Is 'New Conservation' Still Conservation?" *Native Plant News* 1: 6–10.

Farnsworth, E. J., A. A. Barker Plotkin, and A. M. Ellison. 2012. "The Relative Contributions of Seed Bank, Seed Rain, and Understory Vegetation Dynamics to the Reorganization of *Tsuga canadensis* Forests after Loss Due to Logging or Simulated Attack by *Adelges tsugae*." *Canadian Journal of Forest Research* 42: 2090–2105.

Fegraus, E. H., S. Andelman, M. B. Jones, and M. Schildhauer. 2005. "Maximizing the Value of Ecological Data with Structured Metadata: An Introduction to Ecological Metadata Language (EML) and Principles for Metadata Creation." *Bulletin of the Ecological Society of America* 86: 158–168.

Feyerabend, P. 1975. *Against Method.* London: New Left.

Field, S. A., P. J. O'Connor, A. J. Tyre, and H. P. Possingham. 2007. "Making Monitoring Meaningful." *Austral Ecology* 32: 485–491.

Filchak, K. E., J. B. Roethele, and J. L. Feder. 2000. "Natural Selection and Sympatric Divergence in the Apple Maggot *Rhagoletis pomonella*." *Nature* 407: 739–742.

Finkelstein, A., and S. Taubman. 2015. "Randomize Evaluations to Improve Health Care Delivery." *Science* 347: 720–722.

Fischer, H. 1939, 1940. "Geschichte der Kartographie von Palästina." *Zeitschrift des Deutschen Palästina Vereins* 62: 169–189 and 63: 27–29, 43–45.

Fisher, R. A. 1925. *Statistical Methods for Research Workers*. Edinburgh: Oliver and Boyd.

Fitzpatrick, M. C., E. L. Preisser, A. M. Ellison, and J. S. Elkinton. 2009. "Observer Bias and the Detection of Low-Density Populations." *Ecological Applications* 19: 1673–1679.

Fitzpatrick, M. C., N. J. Gotelli, and A. M. Ellison. 2013. "MaxEnt vs. MaxLike: Empirical Comparisons with Ant Species Distributions." *Ecosphere* 4: 55.

Foster, D. R., and D. A. Orwig. 2006. "Pre-emptive and Salvage Harvesting of New England Forests: When Doing Nothing Is a Viable Alternative." *Conservation Biology* 20: 959–970.

Foster, R. G., and L. Kreitzman. 2005. *Rhythms of Life: The Biological Clocks That Control the Daily Lives of Every Living Thing*. New Haven, CT: Yale University Press.

Foucault, M. 1970. *The Order of Things*. New York: Vintage.

Francis, T. I., E. A. Lewis, A. B. Oyediran, O. A. Okubadejo, et al. 1971. "Effect of Chemotherapy on the Duration of Diarrhoea, and on Vibrio Excretion by Cholera Patients." *Journal of Tropical Medicine and Hygiene* 74: 172–176.

Franco, A., N. Malhotra, and G. Simonovits. 2014. "Publication Bias in the Social Sciences: Unlocking the File Drawer. *Science* 345: 1502–1505.

Freeman, B. G., and A. M. Class Freeman. 2014. "Rapid Upslope Shifts in New Guinean Birds Illustrate Strong Distributional Responses of Tropical Montane Species to Global Warming." *Proceedings of the National Academy of Sciences, USA*, 111: 4490–4494.

Gadelha, L.M.R. Jr., M. Mattoso, M. Wilde, and I. Foster. 2011. "Provenance Query Patterns for Many-Task Scientific Computing." In *Proceedings of TaPP '11, 3rd USENIX Workshop on the Theory and Practice of Provenance*, Heraklion, Greece.

Gafter-Gvili, A., A. Fraser, M. Paul, L. Vidal, et al. 2012. "Antibiotic Prophylaxis for Bacterial Infections in Afebrile Neutropenic Patients Following Chemotherapy." *Cochrane Database Systematic Reviews* CD004386.

Galil, B. S., and C. Lewinsohn. 1981. "Macrobenthic Communities of the Eastern Mediterranean Continental Shelf." *PSZNI Marine Ecology* 2: 343–352.

Gardner, J. L., T. Amano, W. J. Sutherland, L. Joseph, et al. 2014. "Are Natural History Collections Coming to an End as a Time-Series?" *Frontiers in Ecology and the Environment* 12: 436–438.

Gaston, K. J., S. L. Chown, and K. L. Evans. 2008. "Ecogeographical Rules: Elements of a Synthesis." *Journal of Biogeography* 35: 483–500.

Gelman, A., and C. R. Shalizi. 2013. "Philosophy and Practice of Bayesian Statistics." *British Journal of Mathematical and Statistical Psychology* 66: 8–38.

Gerson, Elihu M. 2007. "Reach, Bracket, and the Limits of Rationalized Coordination: Some Challenges for CSCW." Pages 193–220 in *Resources, Co-evolution and Artifacts: Theory in CSCW*, edited by W. A. Ackerman, M. S. Halverson, C. A. Erickson, and Th. Kellogg. Dordrecht: Springer-Verlag New York.

——. 2013. "Integration of Specialties: An Institutional and Organizational View." *Studies in History and Philosophy of Biological and Biomedical Sciences* 44: 515–524.

Gharaibeh, S., and K. Mahmoud. 2013. "Decay of Maternal Antibodies in Broiler Chickens." *Poultry Science* 92: 2333–2336.

Gharaibeh, S., K. Mahmoud, and M. Al-Natour. 2008. "Field Evaluation of Maternal Antibody Transfer to a Group of Pathogens in Meat-Type Chickens." *Poultry Science* 87: 1550–1555.

Gharagozloo, R. A., K. Naficy, M. Mouin, M. H. Nassirzadeh, et al. 1970. "Comparative Trial of Tetracycline, Chloramphenicol, and Trimethoprim-Sulphamethoxazole in Eradication of *Vibrio cholerae* El Tor." *British Medical Journal* 4: 281–282.

Gil, I. S., W. Sheldon, T. Schmidt, M. Servilla, et al. 2008. "Defining Linkages between the GSC and NSF's LTER Program: How the Ecological Metadata Language (EML) Relates to GCDML and Other Outcomes." *Omics—A Journal of Integrative Biology* 12: 151–156.

Godlewska, A.M.C. 1999. *Geography Unbound: French Geographic Science from Cassini to Humboldt*. Chicago: University of Chicago Press.

Goren, H. 1994–1995. "Titus Tobler's Legacy: Two Sources." *Bulletin of the Anglo-Israel Archaeological Society* 14: 57–62.

——. 1997. "Nicolayson and Finn Describe the Expeditions and Deaths of Costigan and Molyneux." *Cathedra* 85: 65–94 [in Hebrew and English].

——. 1998. "The Chase after the Bible: Individuals and Institutions—and the Study of the Holy Land." Pages 103–115 in *Religion, Ideology and Geographical Thought*, edited by U. Wardenga and W. J. Wilczynski. Kielce, Poland: Wyzsza Szkola Pedagogiczna im. Jana Kochanowskiego, Instytut Geografii.

——. 1999. "Heinrich Kiepert in the Holy Land, Spring 1870—Sketches from an Exploration-Tour of an Historical Cartographer." Pages 45–62 in *Antike Welten Neue Regionen: Heinrich Kiepert 1818–1899*, edited by L. Zögner. Berlin: Kiepert.

——. 2002. "Sacred, but Not Surveyed: Nineteenth-Century Surveys of Palestine." *Imago Mundi* 54: 87–110.

——. 2003. "Zieht hin und erforscht das Land: Die deutsche Palästinaforschung im 19. Jahrhundert" (translated by A. C. Naujoks). In *Schriftenreihe des Instituts für deutsche Geschichte der Univesität Tel Aviv* 23. Göttingen: Wallstein.

——. 2005. "British Surveyors in Palestine and Syria, 1840–1841." Paper presented at the 2005 *International Cartographic Conference*, Coruña, Spain.

Available at http://www.cartesia.org/geodoc/icc2005/pdf/oral/TEMA16
/Session%205/HAIM%20GOREN.pdf. Accessed 29 June 2016.

———. 2010. "Israeli Scholars since 1970 and the Study of the European Presence
in Palestine in the Nineteenth Century (until World War I): State of the
Art." Pages 55–73 in *Europe and Palestine 1799–1948: Religion—Politics—
Society*, edited by B. Haider-Wilson and D. Trimbur. Vienna: Historische
Kommission, Österreichische Akademie der Wissenshaften.

———. 2011. *Dead Sea Level: Science, Exploration and Imperial Interests in the Near
East*. London: I. B. Tauris.

Goren, H., and B. Schelhaas. 2014. "On an Unknown Measuring of Jerusalem
in 1823." *Die ERDE* 146: 63–78.

Gotelli, N. J., and A. M. Ellison. 2012. *A Primer of Ecological Statistics*, 2nd
edition. Sunderland, MA: Sinauer Associates.

Graham, C. H., S. Ferrier, F. Huettman, C. Moritz, et al. 2004. "New Developments in Museum-Based Informatics and Applications in Biodiversity
Analysis." *Trends in Ecology and Evolution* 19: 497–503.

Grayson, D. K. 2011. *The Great Basin: A Natural Prehistory*. Berkeley: University
of California Press.

Gremillet, D., and T. Boulinier. 2009. "Spatial Ecology and Conservation of
Seabirds Facing Global Climate Change: A Review." *Marine Ecology Progress
Series* 391: 121–137.

Griesemer, J. R. 1990. "Modeling in the Museum: On the Role of Remnant
Models in the Work of Joseph Grinnell." *Biology and Philosophy* 5: 3–36.

Griesemer, J. R., and E. Gerson. 1993. "Collaboration in the Museum of Vertebrate Zoology." *Journal of the History of Biology* 26: 185–203.

Grinnell, J. 1908. Letter to A. Alexander, February 18, 1908. Archives, University
of California, Berkeley.

———. 1910. "The Methods and Uses of a Research Museum." *Popular Science
Monthly* 75: 163–169.

Gropp, R. E. 2003. "Are University Science Collections Going Extinct?" *BioScience* 53: 550 .

Guillera-Arroita, G., J. J. Lahoz-Monfort, D. I. MacKenzie, B. A. Wintle, et al.
2014. "Ignoring Imperfect Detection in Biological Surveys Is Dangerous: A
Response to 'Fitting and Interpreting Occupancy Models.' " *PLOS ONE* 9:
e99571.

Guisan, A., and W. Thuiller. 2005. "Predicting Species Distribution: Offering
More than Simple Habitat Models." *Ecology Letters* 8: 993–1009.

Gurdasani, D., T. Carstensen, F. Tekola-Ayele, L. Pagani, et al. 2015. "The
African Genome Variation Project Shapes Medical Genetics in Africa."
Nature 517: 327–332.

Gwinner, E., and B. Helm. 2003. "Circannual and Circadian Contributions to
the Timing of Avian Migration." Pages 81–95 in *Avian Migration*, edited by
P. Berthold, E. Gwinner, and E. Sonnenschein. Heidelberg: Springer.

Gwinner, E., and W. Wiltschko. 1980. "Circannual Changes in the Migratory Orientation of the Garden Warbler, *Sylvia borin*." *Behavioural Ecology and Sociobiology* 7: 73–78.

Habre, C., M. R. Tramèr, D. M. Pöpping, and N. Elia. 2014. "Ability of a Meta-analysis to Prevent Redundant Research: Systematic Review of Studies on Pain from Propofol Injection." *British Medical Journal* 349: g5219.

Hacking, I. 1983. *Representing and Intervening*. Cambridge: Cambridge University Press.

Hadly, E. A. 1996. "Influence of Late-Holocene Climate on Northern Rocky Mountain Mammals." *Quaternary Research* 46: 298–310.

Halle, S., and N. C. Stenseth, editors. 2000. *Activity Patterns in Small Mammals*. Berlin: Springer.

Halliday, D., R. Resnick, and J. Walker. 2005. *Fundamentals of Physics,* 7th edition. New York: John Wiley and Sons.

Hamal, K. R., S. C. Burgess, I. Y. Pevzner, and G. F. Erf. 2006. "Maternal Antibody Transfer from Dams to Their Egg Yolks, Egg Whites, and Chicks in Meat Lines of Chickens." *Poultry Science* 85: 1364–1372.

Hamilton, W. R. 1842. "Address to the Royal Geographical Society of London: Delivered at the Anniversary Meeting of the 23rd May, 1842." *Journal of the Royal Geographical Society* 12: xxxv–lxxxvii.

Hammers, M., D. S. Richardson, T. Burke, and J. Komdeur. 2012. "Age-Dependent Terminal Declines in Reproductive Output in a Wild Bird." *PLOS ONE* 7: e40413.

Hammond, D. M., A. Adriaanse, E. Rodenburg, D. Bryant, et al. 1995. *Environmental Indicators: A Systematic Approach to Measuring and Reporting on Environmental Policy Performance in the Context of Sustainable Development*. Washington, DC: World Resources Institute.

Hampton, S. E., C. A. Strasser, J. J. Tewksbury, W. K. Gram, et al. 2013. "Big Data and the Future of Ecology." *Frontiers in Ecology and the Environment* 11: 156–162.

Hayden, E. C. 2013. "Weak Statistical Standards Implicated in Scientific Irreproducibility." *Nature News,* 11 November 2013. Available at http://dx.doi.org/10.1038/nature.2013.14131. Accessed 20 June 2016.

Haydon, D. T., N. C. Stenseth, M. S. Boyce, and P. E. Greenwood. 2001. "Phase Coupling and Synchrony in the Spatiotemporal Dynamics of Muskrat and Mink Populations across Canada." *Proceedings of the National Academy of Sciences, USA,* 98: 13149–13154.

Head, J. A., A. DeBofsky, J. Hinshaw, and N. Basu. 2011. "Retrospective Analysis of Mercury Content in Feathers of Birds Collected from the State of Michigan (1895–2007)." *Ecotoxicology* 20: 1636–1643.

Heidinger, B. J., J. D. Blount, W. Boner, K. Griffiths, et al. 2012. "Telomere Length in Early Life Predicts Lifespan." *Proceedings of the National Academy of Sciences, USA,* 109: 1743–1748.

Heink, U., and I. Kowarik. 2010. "What Criteria Should Be Used to Select Biodiversity Indicators?" *Biodiversity Conservation* 19: 3769–3797.

Heller, R., M. Bogomolov, and Y. Benjamini. 2014. "Deciding Whether Follow-up Studies Have Replicated Findings in a Preliminary Large-Scale Omics Study." *Proceedings of the National Academy of Sciences, USA*, 111: 16262–16267.

Helm, B. 2009. "Geographically Distinct Reproductive Schedules in a Changing World: Costly Implications in Captive Stonechats." *Integrative and Comparative Biology* 49: 563–579.

Helm, B., and M. E. Visser. 2010. "Heritable Circadian Period Length in a Wild Bird Population." *Proceedings of the Royal Society B: Biological Sciences*, 277: 3335–3342.

Helm, B., T. Piersma, and H. van der Jeugd. 2006. "Sociable Schedules: Interplay between Avian Seasonal and Social Behaviour." *Animal Behaviour* 72: 245–262; 1215.

Helm, B., E. Gwinner, A. Koolhaas, P. Battley, et al. 2012. "Avian Migration: Temporal Multitasking and a Case Study of Melatonin Cycles in Waders." *Progress in Brain Research* 199: 457–479.

Herman, S. G. 1986. *A Manual of Instruction Based on a System Established by Joseph Grinnell.* Vermillion, SD: Buteo.

Hesse, M. B. 1966. *Models and Analogies in Science,* revised edition. Notre Dame, IN: Notre Dame University Press.

Hobbie, J. E., S. R. Carpenter, N. B. Grimm, J. R. Gosz, et al. 2003. "The US Long Term Ecological Research Program." *BioScience* 53: 21–32.

Hoegh-Guldberg, O., P. J. Mumby, A. J. Hooten, R. S. Steneck, et al. 2007. "Coral Reefs under Rapid Climate Change and Ocean Acidification." *Science* 318: 1737–1742.

Holland, R. A., and B. Helm. 2013. "A Strong Magnetic Pulse Affects the Precision of Departure Direction of Naturally Migrating Adult but Not Juvenile Birds." *Journal of the Royal Society Interface* 10: 2192–2200.

Hollander, M., D. A. Wolfe, and E. Chicken. 2014. *Nonparametric Statistical Methods.* Hoboken, NJ: John Wiley and Sons.

Holling, C. S., editor. 1978. *Adaptive Environmental Assessment and Management.* Chichester, UK: John Wiley and Sons.

Hook, L. A., S. K. Santhana Vannan, T. W. Beaty, R. B. Cook, et al. 2010. "Best Practices for Preparing Environmental Data Sets to Share and Archive." Oak Ridge, TN: Oak Ridge National Laboratories. Available at http://daac.ornl.gov/PI/BestPractices-2010.pdf. Accessed 29 June 2016

Horta, F., J. Dias, R. Elias, D. Oliveira, et al. 2013. "Prov-Vis: Large-Scale Scientific Data Visualization Using Provenance." Poster presented at *SC13: The International Conference for High Performance Computing, Networking, Storage and Analysis,* Denver, Colorado. Available at http://sc13.supercomputing.org/sites/default/files/PostersArchive/post281.html. Accessed 29 June 2016.

Hossain, M. S., M. A. Salam, G. H. Rabbani, I. Kabir, et al. 2002. "Tetracycline in the Treatment of Severe Cholera Due to *Vibrio cholerae* O139 Bengal. *Journal of Health, Population, and Nutrition* 20: 18–25.

Howson, C., and P. Urbach. 1989. *Scientific Reasoning: The Bayesian Approach.* La Salle, IL: Open Court.

Hubisz, M. J., M. F. Lin, M. Kellis, and A. Siepel. 2011. "Error and Error Mitigation in Low-Coverage Genome Assemblies." *PLOS ONE* 6: e17034.

Hueppop, O., and K. Hueppop. 2003. "North Atlantic Oscillation and Timing of Spring Migration in Birds." *Proceedings of the Royal Society B: Biological Sciences,* 270: 233–240.

Humphreys, P. 2007. *Extending Ourselves.* Oxford: Oxford University Press.

Hurlbert, S. H. 1984. "Pseudoreplication and the Design of Ecological Field Experiments." *Ecological Monographs* 54: 187–211.

IEEE (Institute of Electrical and Electronics Engineers) Standard Computer Dictionary. 1990. *A Compilation of IEEE Standard Computer Glossaries.* New York: Institute of Electrical and Electronics Engineers.

Ilves, K. L., A. M. Quattrini, M. W. Westneat, R. I. Eytan, et al. 2013. "Detection of Shifts in Coral Reef Fish Assemblage Structure over 50 Years at Reefs of New Providence Island, The Bahamas, Highlight the Value of the Academy of Natural Sciences' Collections in a Changing World." *Proceedings of the Academy of Natural Sciences of Philadelphia* 162: 61–87.

Ioannidis, J.P.A. 2005. "Why Most Published Research Findings Are False." *PLOS Medicine* 2: e124.

Ioannidis, J.P.A., R. Tarone, and J. McLaughlin. 2011. "The False-Positive to False-Negative Ratio in Epidemiologic Studies." *Epidemiology* 22: 450–456.

Ioannidis, J.P.A., M. R. Munafò, P. Fusar-Poli, B. A. Nosek, et al. 2014. "Publication and other Reporting Biases in Cognitive Sciences: Detection, Prevalence, and Prevention." *Trends in Cognitive Science* 18: 235–241.

Islam, M. R. 1987. "Single Dose Tetracycline in Cholera." *Gut* 28: 1029–1032.

Ives, T. 2013. "Theoretical Inference from Models and Data." *Official Meeting Program of the 98th Annual ESA Meeting: Sustainable Pathways; Learning from the Past and Shaping the Future,* Minneapolis, Minnesota.

Jablonski, D. 2003. "The Interplay of Physical and Biotic Factors in Macroevolution." Pages 235–252 in *Evolution on Planet Earth,* edited by L. J. Rothschild and A. M. Lister. San Diego: Academic Press.

Jackson, J.B.C., M. X. Kirby, W. H. Berger, K. A. Bjorndal, et al. 2001. "Historical Overfishing and the Recent Collapse of Coastal Ecosystems." *Science* 293: 629–637.

Jampoler, A.C.A. 2005. *Sailors in the Holy Land: The 1848 American Expedition to the Dead Sea and the Search for Sodom and Gomorrah.* Annapolis, MD: Naval Institute Press.

Jiao, L. 2009. "China Searches for an 11th Hour Lifesaver for a Dying Discipline." *Science* 325: 31.

Johnson, K. G., S. J. Brooks, P. B. Fenberg, A. G. Glover, et al. 2010. "Climate Change and Biosphere Response: Unlocking the Collections Vault." *BioScience* 61: 147–153.

Jones, H. P., and O. J. Schmitz. 2009. "Rapid Recovery of Damaged Ecosystems." *PLOS ONE* 4: e3653.

Jones, J. P., B. Collen, G. Atkinson, P. W. Baxter, et al. 2011. "The Why, What, and How of Global Biodiversity Indicators beyond the 2010 Target." *Conservation Biology* 25: 450–457.

Jones, M. B., M. P. Schildhauer, O. J. Reichman, and S. Bowers. 2006. "The New Bioinformatics: Integrating Ecological Data from the Gene to the Biosphere." *Annual Review of Ecology, Evolution, and Systematics,* 37: 519–544.

Jones, O. R., A. Scheuerlein, R. Salguero-Gómez, C. G. Camarda, et al. 2014. "Diversity of Ageing across the Tree of Life." *Nature* 505: 169–173.

Jones, Y. 1973. "British Military Surveys of Palestine and Syria 1840–1841." *The Cartographic Journal* 10: 29–41.

Joppa, L. N., D. L. Roberts, and S. L. Pimm. 2011. "The Population Ecology and Social Behavior of Taxonomists." *Trends in Ecology and Evolution* 26: 551–553.

Kabir, I., W. A. Khan, R. Haider, A. K. Mitra, et al. 1996. "Erythromycin and Trimethoprim-Sulphamethoxazole in the Treatment of Cholera in Children." *Journal of Diarrhoeal Disease Research* 14: 243–247.

Kaeffer, B., and L. Pardini. 2005. "Clock Genes of Mammalian Cells: Practical Implications in Tissue Culture." *In Vitro Cellular and Developmental Biology—Animal* 41: 311–320.

Kaiser, T. S. 2014. "Local Adaptations of Circalunar and Circadian Clocks: The Case of *Clunio marinus.*" Pages 121–141 in *Annual, Lunar and Tidal Clocks: Patterns and Mechanisms of Nature's Enigmatic Rhythms,* edited by H. Numata and B. Helm. Tokyo: Springer Japan.

Karchmer, A. W., G. T. Curlin, M. I. Huq, and N. Hirschhorn. 1970. "Furazolidone in Pædiatric Cholera." *Bulletin of the World Health Organization* 43: 373–378.

Kareiva, P. 1989. "Renewing the Dialogue between Theory and Experiments in Population Ecology." Pages 68–88 in *Perspectives in Ecological Theory,* edited by J. Roughgarden, R. M. May, and S. A. Levin. Princeton, NJ: Princeton University Press.

Kareiva, P., and M. Andersen. 1988. "Spatial Aspects of Species Interactions: The Wedding of Models and Experiments." Pages 35–50 in *Community Ecology,* edited by A. Hastings. Berlin: Springer.

Kareiva, P., and M. Marvier. 2012. "What Is Conservation Science?" *BioScience* 62: 962–969.

Karl, D. M., and R. Lukas. 1996. "The Hawaii Ocean Time-series (HOT) Program: Background, Rationale and Field Implementation." *Deep Sea Research II* 43: 129–156.

Kati, V., P. Devillers, M. Dufrêne, A. Legakis, et al. 2004. "Testing the Value of Six Taxonomic Groups as Biodiversity Indicators at a Local Scale. *Conservation Biology* 18: 667–675.

Kemp, C. 2015. "The Endangered Dead." *Nature* 518: 292–294.

Khan, W. A., M. Begum, M. A. Salam, P. K. Bardhan, et al. 1995. "Comparative Trial of Five Antimicrobial Compounds in the Treatment of Cholera in Adults." *Transactions of the Royal Society, Tropical Medicine and Hygiene*, 89: 103–106.

Kiepert, H. 1841. *Plan of Jerusalem: Sketched from Sieber and Catherwood, Corrected by the Measurements of Robinson and Smith.* London: John Murray.

——. 1845. *Plan von Jerusalem nach den Untersuchungen von Dr. Ernst Gustav Schultz, K. Preuss. Consul in Jerusalem.* Berlin: S. Schropp.

Kierkegaard, S. 1843. *Repetitions and Philosophical Crumbs.* Oxford: Oxford University Press, 2009.

Kilkenny, C., W. J. Browne, I. C. Cuthill, M. Emerson, et al. 2010. "Improving Bioscience Research Reporting: The ARRIVE Guidelines for Reporting Animal Research." *PLOS Biology* 8: e1000412.

Kitaoka, M., S. T. Miyata, D. Unterweger, and S. Pukatzki. 2011. "Antibiotic Resistance Mechanisms of *Vibrio cholerae*." *Journal of Medical Microbiology* 60: 397–407.

Klein, C. 1982. "Morphological Evidence of Lake Level Changes, Western Shore of the Dead Sea." *Israel Journal of Earth-Sciences* 31: 67–94.

——. 1986. *Fluctuations of the Level of the Dead Sea and Climatic Fluctuations in Eretz-Israel During Historical Periods.* Ph.D. Dissertation, The Hebrew University, Jerusalem [in Hebrew].

Kohler, R. E. 2012. "Practice and Place in Twentieth Century Field Biology: A Comment." *Journal of the History of Biology* 45: 579–586.

Koop, D., J. Freire, and C. T. Silva. 2013. "Enabling Reproducible Science with VisTrails." Paper presented at The First Workshop on Sustainable Software for Science: Practice and Experiences (WSSSPE1), Denver, Colorado. Available at http://arxiv.org/abs/1309.1784. Accessed 29 June 2016.

Koricheva, J., J. Gurevitch, and K. Mengersen, editors. 2013. *Handbook of Meta-analysis in Ecology and Evolution.* Princeton, NJ: Princeton University Press.

Körner, C. 2003. *Alpine Plant Life: Functional Plant Ecology of High Mountain Ecosystems,* 2nd edition. Berlin: Springer.

Krawetz, S. A. 1989. "Sequence Errors Described in GenBank: A Means to Determine the Accuracy of DNA Sequence Interpretation." *Nucleic Acids Research* 17: 3951–3957.

Kremen, C. 2005. "Managing Ecosystem Services: What Do We Need to Know about Their Ecology?" *Ecology Letters* 8: 468–479.

Kripke, S. A. 1972. "Naming and Necessity." Pages 253–355 in *Semantics of Natural Language,* 2nd edition, edited by D. Davidson and G. Harman. Dordrecht: D. Reidel.

Kruse, F., editor. 1854. *Ulrich Jasper Seetzen's Reisen durch Syrien, Palästina, Phönicien, die Transjordan-Länder, Arabia Petraea und Unter-Aegypten.* Berlin: G. Reimer.

Kruskall, M. S., C. A. Alper, Z. Awdeh, E. J. Yunis, et al. 1992. "The Immune Response to Hepatitis B Vaccine in Humans: Inheritance Patterns in Families." *Journal of Experimental Medicine* 175: 495–502.

Kuhn, T. S. 1962. *The Structure of Scientific Revolutions.* Chicago: University of Chicago Press.

Lagoze, C., J. Willliams, and L. Vilhuber. 2013. "Encoding Provenance Metadata for Social Science Datasets." Pages 123–134 in *Metadata and Semantics Research,* edited by E. Garoufallou, and J. Greenberg. Cham, Switzerland: Springer International.

Lane, J. E., L.E.B. Kruuk, A. Charmantier, J. O. Murie, et al. 2012. "Delayed Phenology and Reduced Fitness Associated with Climate Change in a Wild Hibernator." *Nature* 489: 554–557.

Laney, C. M., and D.P.C. Peters. 2006. "Trends in Long-Term Ecological Data: A Collaborative Synthesis Project Introduction and Update." *Databits.* Available at http://databits.lternet.edu/spring-2006/trends-long-term -ecological-data-collaborative-synthesis-project-introduction-and-update. Accessed 29 June 2016.

Lapeysonnie, L., H. Aye, J. Rive, M. Duchassin, et al. 1971. "Controlled Trials on Place of Sulformethoxine on the Carrying of Choleric Vibrions." *La Presse Médicale* 79: 637–638. [in French].

Larigauderie, A., and H. A. Mooney. 2010. "The International Year of Biodiversity: An Opportunity to Strengthen the Science-Policy Interface for Biodiversity and Ecosystem Services." *Current Opinion in Environmental Sustainability* 2: 1–2.

Lawton, J. H. 1999. "Are There General Laws in Ecology?" *Oikos* 84: 177–192.

Leek, J. T., and R. D. Peng. 2015. "What Is the Question?" *Science* 347: 1314–1315.

Leek, J. T., R. B. Scharpf, H. C. Bravo, D. Simcha, et al. 2010. "Tackling the Widespread and Critical Impact of Batch Effects in High-Throughput Data." *Nature Reviews: Genetics* 11: 733–739.

Legg, C. J., and L. Nagy. 2006. "Why Most Conservation Monitoring Is, but Need Not Be, a Waste of Time." *Journal of Environmental Management* 78: 194–199.

Leibovici-Weissman, Y., A. Neuberger, R. Bitterman, D. Sinclair, et al. 2014. "Antimicrobial Drugs for Treating Cholera." *Cochrane Database Systematic Reviews* CD008625.

Lemmens, L. 1916. *Die Franziskaner im Hl. Lande, I: Die Franziskaner auf dem Sion (1336–1551).* Munster, Germany: Aschendorffsche.

Leonelli, S. 2013. "Integrating Data to Acquire New Knowledge: Three Modes of Integration in Plant Science." *Studies in History and Philosophy of Science, Part C: Studies in History and Philosophy of Biological and Biomedical Sciences* 44, no. 4, Part A: 503–514.

Lerner, B. S., and E. R. Boose. 2014a. "RDataTracker and DDG Explorer: Capture, Visualization and Querying of Provenance from R Scripts." Poster presented at IPAW 2014, the International Provenance and Annotation Workshop, Cologne, Germany.

———. 2014b. "RDataTracker: Collecting Provenance in an Interactive Scripting Environment." In *Proceedings of the 6th USENIX Workshop on the Theory and Practice of Provenance* (TaPP '14), Cologne, Germany. Berkeley, CA: USENIX Association. Available at https://www.usenix.org/system/files/conference /tapp2014/tapp14_paper_lerner.pdf. Accessed 29 June 2016.

Lerner, B., E. Boose, L. J. Osterweil, A. M. Ellison, et al. 2011. Provenance and "Quality Control in Sensor Networks." Pages 98–103 in *EIM 2011: Proceedings of the Environmental Information Management Conference*, edited by M. B. Jones and C. Gries. Santa Barbara, CA.

Levins, R. 1966. "The Strategy of Model Building in Population Biology." Pages 18–27 in *Conceptual Issues in Evolutionary Biology*, edited by E. Sober. Cambridge, MA: MIT Press.

———. 1968. *Evolution in Changing Environments*. Princeton, NJ: Princeton University Press.

Lindenbaum, J., W. B. Greenough, and M. R. Islam. 1967a. "Antibiotic Therapy of Cholera." *Bulletin of the World Health Organization* 36: 871–883.

———. 1967b. "Antibiotic Therapy of Cholera in Children." *Bulletin of the World Health Organization* 37: 529–538.

Lindenmayer, D. B., and G. E. Likens. 2009. "Adaptive Monitoring: A New Paradigm for Long Term Research and Monitoring." *Trends in Ecology and Evolution* 24: 482–487.

———. 2010a. *Effective Ecological Monitoring*. Collingwood, Australia: CSIRO.

———. 2010b. "The Science and Application of Ecological Monitoring." *Biological Conservation* 143: 1317–1328.

———. 2011. "Direct Measurements Versus Surrogate Indicator Species for Evaluating Environmental Change and Biodiversity Loss." *Ecosystems* 14: 47–59.

Lindenmayer, D. B., A. D. Manning, P. L. Smith, H. P. Possingham, et al. 2002. "The Focal-Species Approach and Landscape Restoration: A Critique." *Conservation Biology* 16: 338–345.

Lister, A. M., and the Climate Change Research Group. 2011. "Natural History Collections as Sources of Long-Term Data Sets." *Trends in Ecology and Evolution* 26: 153–154.

Livio, M. 2014. *Brilliant Blunders: From Darwin to Einstein—Colossal Mistakes by Great Scientists That Changed Our Understanding of Life and the Universe.* New York: Simon and Schuster.

Lloyd, E. A. 1988. *The Structure and Confirmation of Evolutionary Theory.* Westport, CT: Greenwood.

Lolekha, S., S. Patanachareon, B. Thanangkul, and S. Vibulbandhitkit. 1988. "Norfloxacin Versus Co-Trimoxazole in the Treatment of Acute Bacterial

Diarrhoea: A Placebo Controlled Study." *Scandinavian Journal of Infectious Disease Supplement* 56: 35–45.

Lotze, H., and B. Worm. 2009. "Historical Baselines for Large Marine Mammals." *Trends in Ecology and Evolution* 24: 254–262.

Loya, Y. 2007. "How to Influence Environmental Decision Makers? The Case of Eilat (Red Sea) Coral Reefs." *Journal of Experimental Marine Biology and Ecology* 344: 35–53.

Loya, Y., H. Lubinevsky, M. Rosenfeld, and W. E. Kramarsky. 2004a. "Nutrient Enrichment Caused by in Situ Fish Farms at Eilat, Red Sea, Is Detrimental to Coral Reproduction." *Marine Pollution Bulletin* 49: 344–353.

———. 2004b. "Nutrient Enrichment and Coral Reproduction: Empty Vessels Make the Most Sound (Response to Critique by B. Rinkevich)." *Marine Pollution Bulletin* 50: 114–118.

Ludäscher, B., I. Altintas, C. Berkley, D. Higgins, et al. 2006. "Scientific Workflow Management and the Kepler System." *Concurrency and Computation: Practice and Experience* 18: 1039–1065.

Lynch, W. F. 1849. *Narrative of the United States' Expedition to the River Jordan and the Dead Sea*. Philadelphia: Lea and Blanchard.

———. 1852. *Official Report of the United States' Expedition to Explore the Dead Sea and the River Jordan*. Baltimore: John Murphy.

Maas, R. A., M. Komen, M. van-Diepen, H. L. Oei, et al. 2003. "Correlation of Haemagglutinin-Neuraminidase and Fusion Protein Content with Protective Antibody Response after Immunisation with Inactivated Newcastle Disease Vaccines." *Vaccine* 21: 3137–3142.

MacCoun, R., and S. Perlmutter. 2015. "Hide Results to Seek the Truth." *Nature* 526: 187–189.

MacKenzie, D. I., J. D. Nichols, G. B. Lachman, S. Droege, et al. 2002. "Estimating Site Occupancy Rates When Detection Probabilities Are Less than One." *Ecology* 83: 2248–2255.

MacKenzie, D. I., J. D. Nichols, J. A. Royle, K. H. Pollock, et al. 2006. *Occupancy Estimation and Modeling*. Burlington, VT: Academic Press.

Mackenzie-Graham, A. J., A. Payan, I. D. Dinov, J. D. Horn, et al. 2008. "Neuroimaging Data Provenance Using the LONI Pipeline Workflow Environment." Pages 208–220 in *Provenance and Annotation of Data and Processes*, edited by J. Freire, D. Koop, and L. Moreau. Berlin: Springer.

Magill, A. H., J. D. Aber, W. S. Currie, K. J. Nadelhoffer, et al. 2004. "Ecosystem Response to 15 Years of Chronic Nitrogen Additions at the Harvard Forest LTER, Massachusetts, USA." *Forest Ecology and Management* 196: 7–28.

Mandelik, Y., U. Roll, and A. Fleischer. 2010. "Cost-efficiency of Biodiversity Indicators for Mediterranean Ecosystems and the Effects of Socio-economic Factors." *Journal of Applied Ecology* 47: 1179–1188.

Marvier, M. 2012. "The Value of Nature Revisited." *Frontiers in Ecology and the Environment* 10: 227.

Maynard, L. W. 1910. "President Roosevelt's List of Birds Seen in the White House Grounds and About Washington during His Administration." *Bird-Lore* 12: 53–55.

Mayr, E. 1959. "Agassiz, Darwin, and Evolution." *Harvard Library Bulletin* 13:165–194.

McCullagh, P., and J. A. Nelder. 1989. *Generalized Linear Models,* 2nd edition. London: Chapman and Hall.

McDowell, M. A., C. D. Fryar, C. L. Ogden, and K. M. Flegal. 2008. "Anthropometric Reference Data for Children and Adults: United States, 2003–2006." *National Health Statistics Reports* 10. Available at http://www.cdc.gov/nchs /data/nhsr/nhsr010.pdf. Accessed 24 March 2014.

McGarigal, K., and Marks, B. J. 1995. *FRAGSTATS: Spatial Pattern Analysis Program for Quantifying Landscape Structure.* USDA Forest Service General Technical Report PNW-351.

McGeoch, M. A. 1998. "The Selection, Testing and Application of Terrestrial Insects as Bioindicators." *Biological Reviews* 73: 181–201.

McGeoch, M. A., M. Dopolo, P., Novellie, H. Hendriks, et al. 2011, "A strategic framework for biodiversity monitoring in South African National Parks." *Koedoe* 53: Article 99. Available online: http://dx.doi.org/10.4102/koedoe .v53i2.991. Accessed 3 July 2016.

McNutt, M. 2014a. "Reproducibility." *Science* 343: 229.

———. 2014b. "Journals Unite for Reproducibility." *Science* 346: 679.

McNutt, M., K. Lehnert, B. Hanson, B. A. Nosek, A. M. Ellison, and J. L. King. 2016. "Moving beyond 'Data/Samples Available upon Request' in the Field Sciences." *Science* 351: 1024–1026.

McPhee, J. 1981. *Basin and Range.* New York: Farrar, Straus and Giroux.

Medebielle, P. 1963. *The Diocese of the Latin Patriarchate of Jerusalem,* translated by E. Friedman. Jerusalem.

Meiri, S., D. Guy, T. Dayan, and D. Simberloff. 2009. "Global Change and Carnivore Body Size: Data Are Stasis." *Global Ecology and Biogeography* 18: 240–247.

Michener, W. K., and M. B. Jones. 2012. "Econinformatics: Supporting Ecology as a Data-Intensive Science." *Trends in Ecology and Evolution* 27: 85–92.

Michener, W. K., J. Brunt, J. Helly, T. Kirchner, et al. 1997. "Nongeospatial Metadata for the Ecological Sciences." *Ecological Applications* 7: 330–342.

Milliken, G. A., and D. E. Johnson. 1989. *Analysis of Messy Data,* volume 2: *Nonreplicated Experiments.* New York: Van Nostrand Reinhold.

Milo, G., E. A. Katchman, M. Paul, T. Christiaens, et al. 2005. "Duration of Antibacterial Treatment for Uncomplicated Urinary Tract Infection in Women." *Cochrane Database Systematic Reviews* CD004682.

Minteer, B. A. 2012. *Refounding Environmental Ethics: Pragmatism, Principle, and Practice.* Philadelphia: Temple University Press.

Minteer, B. A., and J. P. Collins. 2008. "From Environmental to Ecological Ethics: Toward a Practical Ethics for Ecologists and Conservationists." *Science and Engineering Ethics* 14: 483–501.

Minteer, B. A., and S. J. Pyne, editors. 2015. *After Preservation: Saving American Nature in the Age of Humans*. Chicago: University of Chicago Press.

Missier, P., K. Belhajjame, J. Zhao, M. Roos, et al. 2008. "Data Lineage Model for Taverna Workflows with Lightweight Annotation Requirements. Pages 17–30 in *Provenance and Annotation of Data and Processes*, edited by J. Freire, D. Koop, and L. Moreau. Berlin: Springer.

Mizutani, Y., N. Tomita, Y. Niizuma, and K. Yoda. 2013. "Environmental Perturbations Influence Telomere Dynamics in Long-Lived Birds in Their Natural Habitat." *Biology Letters* 9: 20130511.

Moore, G. H., and W. G. Beek. 1837. "On the Dead Sea and Some Positions in Syria." *Journal of the Royal Geographical Society* 7: 456.

Moore, I. T., F. Bonier, and J. C. Wingfield. 2005. "Reproductive Asynchrony and Population Divergence between Two Tropical Bird Populations." *Behavioral Ecology* 16: 755–762.

Moore, M. G. 1913. *An Irish Gentleman George Henry Moore: His Travel, His Racing, His Politics*. London: T. Werner Laurie.

Moratelli, R. 2014. "Wildlife Biologists Are on the Right Track: A Mammalogist's View of Specimen Collection." *Zoologia* 31: 413–417.

Moritz, C., J. L. Patton, C. J. Conroy, J. L. Parra, et al. 2008. "Impact of a Century of Climate Change on Small-Mammal Communities in Yosemite National Park, USA." *Science* 322: 261–264.

Moscrop, J. J. 2000. *Measuring Jerusalem: The Palestine Exploration Fund and British Interests in the Holy Land*. London: Leicester University Press.

Müller, F. 2005. "Indicating Ecosystem and Landscape Organization." *Ecological Indicators* 5: 280–294.

Müller, F., C. Baessler, H. Schubert, and S. Klotz. 2010. *Long Term Ecological Research*. Dordrecht: Springer.

Nachman, M. W. 2013. "Genomics and Museum Specimens." *Molecular Ecology* 22: 5966–5968.

Neumann, P. G. 1994. *Computer-Related Risks*. Reading: Addison-Wesley.

Nichols, J. D., and B. K. Williams. 2006. "Monitoring for Conservation." *Trends in Ecology and Evolution* 21: 668–673.

Niemeijer, N., and R. S. de Groot. 2008. "A Conceptual Framework for Selecting Environmental Indicator Sets." *Ecological Indicators* 8: 14–25.

Nietzsche, F. 1874. *On the Uses and Disadvantages of History for Life, Untimely Meditations*. Cambridge: Cambridge University Press, 1997.

———. 1977. *The Portable Nietzsche*. New York: Penguin.

NISO (National Information Standards Organization). 2004. *Understanding Metadata*. Bethesda, MD: NISO.

Nosek, B. A., G. Alter, G. C. Banks, et al. 2015. "Promoting an Open Research Culture." *Science* 348: 1422.

NRC (US National Research Council). 1995. *Finding the Forest in the Trees: The Challenge of Combining Diverse Environmental Data*. Washington, DC: National Academy Press.

———. 2003. *Sharing Publication-Related Data and Materials: Responsibilities of Authorship in the Life Sciences*. Washington DC: National Academy Press.

———. 2005. *The Geological Record of Ecological Dynamics*. Washington DC: National Academies Press.

Numata, H., and B. Helm, editors. 2014. *Annual, Lunar and Tidal Clocks: Patterns and Mechanisms of Nature's Enigmatic Rhythms*. Tokyo: Springer Japan.

O'Brien, T. G., J.E.M. Baillie, L. Krueger, and M. Cuke. 2010. "The Wildlife Picture Index: Monitoring Top Trophic Levels." *Animal Conservation* 13: 335–343.

Obrist, M. K., and P. Duelli. 2010. "Rapid Biodiversity Assessment of Arthropods for Monitoring Average Local Species Richness and Related Ecosystem Services." *Biodiversity and Conservation* 19: 2201–2220.

O'Neill, J. S., and K. A. Feeney. 2014. "Circadian Redox and Metabolic Oscillations in Mammalian Systems." *Antioxidants and Redox Signaling* 20: 2966–2981.

Orwig, D. A., A. A. Barker Plotkin, E. A. Davidson, H. Lux, et al. 2013. "Foundation Species Loss Affects Vegetation Structure More than Ecosystem Function in a Northeastern USA Forest." *PeerJ* 1: e41.

Orzack, S. H., and E. Sober. 1993. "A Critical Assessment of Levins's 'The Strategy of Model Buildling in Population Biology' (1966)." *Quarterly Review of Biology* 68: 533–546.

Osterweil, L. J., A. Wise, L. A. Clarke, A. M. Ellison, et al. 2005. "Process Technology to Facilitate the Conduct of Science." *Lecture Notes in Computer Science* 3840: 403–415.

Osterweil, L. J., L. A. Clarke, A. M. Ellison, E. Boose, et al. 2010. "Clear and Precise Specification of Ecological Data Management Processes and Dataset Provenance." *IEEE Transactions on Automation Science and Engineering* 7: 189–195.

Oyama, S. 1985. *The Ontogeny of Information: Developmental Systems and Evolution*. Durham, NC: Duke University Press.

Ozgo, M., and M. Schilthuizen. 2012. "Evolutionary Change in *Cepaea nemoralis* Shell Colour over 43 Years." *Global Change Biology* 18: 74–81.

Pandolfi, J. M., R. H. Bradbury, E. Sala, T. P. Hughes, et al. 2003. "Global Trajectories of the Long-Term Decline of Coral Reef Ecosystems." *Science* 301: 955–958.

Parkes, D., G. Newell, and D. Cheal, D. 2003. "Assessing the Quality of Native Vegetation: The 'Habitat Hectares' Approach." *Ecological Management and Restoration* 4: 29–38.

Patterson, D. J., J. Cooper, P. M. Kirk, R. L. Pyle, et al. 2010. "Names Are Key to the Big New Biology." *Trends in Ecology and Evolution* 1297: 1–6.

Paul, M., I. Benuri-Silbiger, K. Soares-Weiser, and L. Leibovici. 2004. "Beta Lactam Monotherapy versus Beta Lactam-Aminoglycoside Combination Therapy for Sepsis in Immunocompetent Patients: Systematic Review and Meta-Analysis of Randomised Trials." *British Medical Journal* 328: 668.

Paul, M., A. Lador, S. Grozinsky-Glasberg, and L. Leibovici. 2014. Beta Lactam Antibiotic Monotherapy versus Beta Lactam-Aminoglycoside Antibiotic Combination Therapy for Sepsis. *Cochrane Database Systematic Reviews* CD003344.

Pearson, D. L., A. L. Hamilton, and T. L. Erwin. 2011. "Recovery Plan for the Endangered Taxonomy Profession." *BioScience* 61: 58–63.

Pennisi, E. 2006. "Shaking the Dust Off Agassiz's Museum." *Science* 313: 754–755.

Peters, D.P.C., C. M. Laney, A. E. Lugo, S. L. Collins, et al. 2013. *Long-Term Trends in Ecological Systems: A Basis for Understanding Responses to Global Change.* Washington, DC: United States Department of Agriculture— Agricultural Research Service, Technical Bulletin 1931.

Petroski, H. 1985. *To Engineer Is Human: The Role of Failure in Successful Design.* New York: St. Martin's.

Picard, P., and M. R. Tramèr. 2000. "Prevention of Pain on Injection with Propofol: A Quantitative Systematic Review." *Anesthesia and Analgesia* 90: 963–969.

Pierce, N. F., J. G. Banwell, R. C. Mitra, G. J. Caranasos, et al. 1968. "Controlled Comparison of Tetracycline and Furazolidone in Cholera." *British Medical Journal* 3: 277–280.

Pierotti, E. 1860. *Plan de Jérusalem Ancienne et Moderne.* Paris: Kreppelin.

Pimentel, D., R. Zuniga, and D. Morrison. 2005. "Update on the Environmental and Economic Costs Associated with Alien-Invasive Species in the United States." *Ecological Economics* 52: 273–288.

Platt, J. R. 1964. "Strong Inference." *Science* 146: 347–353.

Ponder, W. F., G. A. Carter, P. Flemons, and R. R. Chapman. 2001. "Evaluation of Museum Collection Data for Use in Biodiversity Assessment." *Conservation Biology* 15: 648–657.

Popper, K. 1959. *The Logic of Scientific Discovery.* London: Routledge.

Por, F. D. 1978. "Lessepsian Migration—The Influx of Red Sea Biota into the Mediterranean by Way of the Suez Canal." *Ecological Studies* 23: 1–238.

Porter, J., P. Arzberger, H.-W. Braun, P. Bryant, et al. 2005. "Wireless Sensor Networks for Ecology." *BioScience* 55: 561–572.

Pyke, G. H., and P. R. Ehrlich. 2010. "Biological Collections and Ecological/ Environmental Research: A Review, Some Observations and a View to the Future." *Biological Reviews* 85: 247–266.

Quine, W.V.O. 1960. *Word and Object.* Cambridge, MA: MIT Press.

Quiroga, R. Q. 2012. "Concept Cells: The Building Blocks of Declarative Memory Functions." *Nature Reviews Neuroscience* 13: 587–597.

Rabbani, G. H., M. R. Islam, T. Butler, M. Shahrier, et al. 1989. "Single-Dose Treatment of Cholera with Furazolidone or Tetracycline in a Double-Blind Randomized Trial." *Antimicrobial Agents and Chemotherapy* 33: 1447–1450.

Rabbani, G. H., T. Butler, M. Shahrier, R. Mazumdar, et al. 1991. "Efficacy of a Single Dose of Furazolidone for Treatment of Cholera in Children." *Antimicrobial Agents and Chemotherapy* 35: 1864–1867.

Raess, M. 2008. "Continental Efforts: Migration Speed in Spring and Autumn in an Inner-Asian Migrant." *Journal of Avian Biology* 39: 13–18.

Rahaman, M. M., M. A. Majid, A. Akmj, and M. R. Islam. 1976. "Effects of Doxycycline in Actively Purging Cholera Patients: A Double-Blind Clinical Trial." *Antimicrobial Agents and Chemotherapy* 10: 610–612.

Raven, P. H. 2003. "Biodiversity and the Future." *American Scientist* 91:1.

R Core Team. 2015. *R: A Language and Environment for Statistical Computing.* Vienna: R Foundation for Statistical Computing.

Reardon, S. 2015. "NIH Disclosure Rules Falter." *Nature* 525: 300–301.

Remsen, J. V. Jr. 1977. "On Taking Field Notes." *The National Audubon Society* 31: 946–953.

Rheinberger, H.-J. 1997. *Toward a History of Epistemic Things: Synthesizing Proteins in the Test Tube.* Stanford, CA: Stanford University Press.

Rickart, E. A. 2001. "Elevational Diversity Gradients, Biogeography and the Structure of Montane Mammal Communities in the Intermountain Region of North America." *Global Ecology and Biogeography* 10: 77–100.

Ringold, P. L., J. Alegria, R. L. Czaplewski, B.S.T. Tolle, et al. 1996. "Adaptive Monitoring Design for Ecosystem Management." *Ecological Applications* 6: 745–747.

Rinkevich, B. 2005a. "Nutrient Entrenchment and Coral Reproduction: Between Truth and Repose (a Critique of Loya et al)." *Marine Pollution Bulletin* 50: 111–113.

——. 2005b. "What Do We Know about Eilat (Red Sea) Reef Degradation? A Critical Examination of the Published Literature." *Journal of Experimental Marine Biology and Ecology* 327: 183–200.

Ritter, C. 1846. "Des kön. Preuss. General-Consuls und Majors Hrn. L. v. Wildenbruch Profilzeichnungen nach barometrischen Nivellements in Syrien." *Monatsberichte über die Verhandlungen der Gesellschaft für Erdkunde zu Berlin* 3: 270–272.

——. 1850. *Vergleichende Erdkunde der Sinai-Halbinsel, von Palästina und Syrien,* volume 2, part 1. Berlin: G. Reimer.

Robinson, T. M. 1987. *Heraclitus: Fragments.* Toronto: University of Toronto Press.

Rocha, L. A., A. Aleixo, G. Allen, F. Almeda, et al. 2014. "Specimen Collection: An Essential Tool." *Science* 344: 814–815.

Rodríguez Caicedo, D., J. M. Cotes Torres, and J. R. Cure. 2012. "Comparison of Eight Degree-Days Estimation Methods in Four Agroecological Regions in Colombia." *Bragantia* 71: 299–307.

Roenneberg, T., A. Wirz-Justice, and M. Merrow. 2003. "Life between Clocks: Daily Temporal Patterns of Human Chronotypes." *Journal of Biological Rhythms* 18: 80–90.

Röhricht, R. 1890 and 1963. *Bibliotheca geographica Palaestinae: Chronologisches Verzeichnis der von 333 bis 1878 verfassten Literatur über das heilige Land.* Berlin (1890) and Jerusalem (1963).

Rosenberg, A. A., L. M. Branscomb, V. Eady, P. C. Frumhoff, et al. 2015. "Congress's Attacks on Science-Based Rules." *Science* 348: 964–966.

Rowe, K. C., K.M.C. Rowe, M. W. Tingley, M. S. Koo, et al. 2014. "Spatially Heterogeneous Impact of Climate Change on Small Mammals of Montane California." *Proceedings of the Royal Society of London B: Biological Sciences,* 282. doi:10.1098/rspb.2014.1857.

Rowe, R. J. 2005. "Elevational Gradient Analyses and the Use of Historical Museum Specimens: A Cautionary Tale." *Journal of Biogeography* 32: 1883–1897.

———. 2007. "Legacies of Land Use and Recent Climatic Change: The Small Mammal Fauna in the Mountains of Central Utah." *American Naturalist* 170: 242–257.

Rowe, R. J., and S. Lidgard. 2009. "Elevational Gradients and Species Richness: Do Methods Change Pattern Perception?" *Global Ecology and Biogeography* 18: 163–177.

Rowe, R. J., and R. C. Terry. 2014. "Small Mammal Responses to Environmental Change: Integrating Past and Present Dynamics." *Journal of Mammalogy* 95: 1157–1174.

Rowe, R. J., J. A. Finarelli, and E. A. Rickart. 2010. "Range Dynamics of Small Mammals along an Elevational Gradient over an 80-year Interval." *Global Change Biology* 16: 2930–2943.

Rowe, R. J., R. C. Terry, and E. A. Rickart. 2011. "Environmental Change and Declining Resource Availability for Small Mammal Communities in the Great Basin." *Ecology* 92: 1366–1375.

Roy, S. K., A. Islam, R. Ali, K. E. Islam, et al. 1998. "A Randomized Clinical Trial to Compare the Efficacy of Erythromycin, Ampicillin and Tetracycline for the Treatment of Cholera in Children." *Transactions of the Royal Society, Tropical Medicine and Hygiene,* 92: 460–462.

Rubin, R. 1989. "The de Angelis Map of Jerusalem (1578) and Its Copies." *Cathedra* 52: 100–111 [in Hebrew].

———. 1990. "The Map of Jerusalem (1538) by Hermannus Borculus and its Copies: A Carto- genealogical Study." *Cartographic Journal* 27: 31–39.

———. 1999. *Image and Reality: Jerusalem in Maps and Views.* Jerusalem: Magnes.

Rüegg, J., C. Gries, B. Bond-Lamberty, G. J. Bowen, et al. 2014. "Completing the Data Life Cycle: Using Information Management in Macrosystems Ecology Research." *Frontiers in Ecology and the Environment* 12: 24–30.

Runnalls, A., and C. Silles. 2012. "Provenance Tracking in R." Pages 237–239 in *Provenance and Annotation of Data and Processes,* edited by P. Groth, and J. Frew. Berlin: Springer.

Russell, J. F. 2013. "If a Job Is Worth Doing, It Is Worth Doing Twice." *Nature* 496: 7.

Ruxton, G. D., and N. Colegrave. 2010. *Experimental Design for the Life Sciences.* Oxford: Oxford University Press.

Sack, D. A., R. B. Sack, G. B. Nair, and A. K. Siddique. 2004. "Cholera." *Lancet* 363: 223–233.

Sætersdal, M., and I. Gjerde. 2011. "Prioritizing Conservation Areas Using Species Surrogate Measures: Consistent with Ecological Theory?" *Journal of Applied Ecology* 48: 1236–1240.

Sanderson, K. 2013. "Bloggers Put Chemical Reactions Through the Replication Mill." *Nature News,* 21 January 2013. Available at http://www.nature.com/news/bloggers-put-chemical-reactions-through-the-replication-mill-1.12262. Accessed 29 June 2016.

Sandford, S. A. 1995. "Apples and Oranges—A Comparison." *Annals of Improbable Research* 1: unpaginated. Available at http://www.improbable.com/airchives/paperair/volume1/v1i3/air-1-3-apples.html. Accessed 29 June 2016.

Sandve, G. K., A. Nekrutenko, J. Taylor, and E. Hovig. 2013. "Ten Simple Rules for Reproducible Computation Research." *PLOS Computational Biology* 9: e1003285.

Sarmento, P. B., J. Cruz, C. Eira, and C. Fonseca. 2011. "Modeling the Occupancy of Sympatric Carnivorans in a Mediterranean Ecosystem." *European Journal of Wildlife Research* 57: 119–131.

Saupe, E. E., J. R. Hendricks, A. Townsend Peterson, and B. S. Lieberman. 2014. "Climate Change and Marine Molluscs of the Western North Atlantic: Future Prospects and Perils." *Journal of Biogeography* 41: 1352–1366.

Savage, L. J. 1954. *The Foundations of Statistics.* New York: John Wiley and Sons.

Schattner, J. 1951. *The Map of Eretz-Israel and Its History.* Jerusalem: Bialik Institute [in Hebrew].

Schickore, J. 2010. "Trying Again and Again: Multiple Repetitions in Early Modern Reports of Experiments on Snake Bites." *Early Science and Medicine* 15: 567–617.

Schienerl, J. 2000. *Der Weg in den Orient. Der Forscher Ulrich Jasper Seetzen: Von Jever in den Jemen (1802–1811).* Oldenburg, Germany: Isenee.

Schmeller, D. S. 2008. "European Species and Habitat Monitoring: Where Are We Now?" *Biodiversity and Conservation* 17: 3321–3326.

Schmidt, K., and J. M. Starck. 2010. "Developmental Plasticity, Modularity, and Heterochrony during the Phylotypic Stage of the Zebra Fish, *Danio rerio.*" *Journal of Experimental Zoology Part B: Molecular and Developmental Evolution,* 314B: 166–178.

Schnalke, T. 2011. "Out of the Cellar." *Nature* 471: 576–577.

Schrödinger, E. 1935. "Die gegenwärtige Situation in der Quantenmechanik." *Naturwissenschaften* 23: 807–812, 823–828, 844–849.

Seetzen, U. J. 1810. *Charte von Palästina reducirt aus den von dem Herrn D. Seetzen an Ort und Stelle entworfenen Handzeichnungen.* Gotha, Germany: Justus Perthes.

Shachak, M., B. Boeken, E. Groner, R. Kadmon, et al. 2008. "Woody Species as Landscape Modulators and Their Effect on Biodiversity Patterns." *Bioscience* 58: 209–221.

Shackelford, C., R. Brown, and R. Conner. 2000. "Red-Bellied Woodpecker (*Melanerpes carolinus*)." Pages 1–24 in *The Birds of North America,* vol. 5, edited by A. Poole and F. Gill. Philadelphia: Birds of North America.

Shaffer, H. B., R. N. Fisher, and C. Davidson. 1998. "The Role of Natural History Collections in Documenting Species Declines." *Trends in Ecology and Evolution* 13: 27–30.

Shapin, S., and S. Schaffer. 1985. *Leviathan and the Air-Pump: Hobbes, Boyle, and the Experimental Life.* Princeton, NJ: Princeton University Press.

Shavit, A. 2014. "You Can't Go Home Again—or Can You? 'Replication' Indeterminacy and 'Location' Incommensurability in Three Biological Re-surveys." Paper presented at the 2014 Biennial Meeting of the Philosophy of Science Association, Chicago. Available at http://philsci-archive.pitt .edu/10876/. Accessed 29 June 2016.

——. 2016. "'Location' Incommensurability and 'Replication' Indeterminacy: Clarifying an Entrenched Conflation by Using an Involved Approach." *Perspectives on Science* 24: 425–442.

Shavit, A., and A. M. Ellison. 2013. "Symposium 17: There and Back Again; Replication Standards in Long-Term Research, Integrating the Field and Database Perspectives for Future Management." *Ecological Society of America Bulletin* 94: 395–397.

Shavit, A., and J. R. Griesemer. 2009. "There and Back; or, the Problem of Locality in Biodiversity Research." *Philosophy of Science* 76: 273–294.

——. 2011a. "Mind the Gaps: Why Are Niche Construction Processes So Rarely Used?" Pages 307–317 in *Lamarckian Transformations,* edited by S. Gisis and E. Jablonka. Cambridge, MA: MIT Press.

——. 2011b. "Transforming Objects into Data: How Minute Technicalities of Recording Species Location Entrench a Basic Theoretical Challenge for Biodiversity." Pages 169–193 in *Science in the Context of Application,* edited by M. Carrier and A. Nordmann. Bielefeld, Germany: Zentrum für interdisziplinäre Forschung.

Shmida, A., and G. Pollak. 2007. *Red Data Book: Endangered Plants in Israel.* Jerusalem: Israel Nature and Parks Authority [in Hebrew].

Shrader-Frechette, K. S., and E. D. McCoy. 1993. *Methods in Ecology: Strategies for Conservation.* Cambridge: Cambridge University Press.

Sieber, F. W. 1818. *Karte von Jerusalem und seine nächsten Umgebungen, geometrisch aufgenommen.* Prague.

Silberman, N. A. 1982. *Digging for God and Country: Exploration, Archeology, and the Secret Struggle for the Holy Land 1799–1917.* New York: Random House.

Silva, A., M. Bacci Jr., C. Gomes De Siqueira, O. Correa Bueno, et al. 2003. "Survival of *Atta sexdens* Workers on Different Food Sources." *Journal of Insect Physiology* 49: 307–313.

Simberloff, D. 1998. "Flagships, Umbrellas, and Keystones: Is Single-Species Management Passé in the Landscape Era?" *Biological Conservation* 83: 247–257.

——. 2013. *Invasive Species—What Everyone Needs to Know.* New York: Oxford University Press.

Simmhan, Y., R. Barga, C. van Ingen., E. Lazowska, et al. 2008. "On Building Scientific Workflow Systems for Data Management in the Cloud." Pages 434–435 in *Proceedings of the 2008 Fourth IEEE International Conference on eScience.* New York: IEEE.

Siontis, K. C., T. Hernandez-Boussard, and J. P. Ioannidis. 2013. "Overlapping Meta-analyses on the Same Topic: Survey of Published Studies." *British Medical Journal* 347: f4501.

Smith, M. D, A. K. Knapp, and S. L. Collins. 2009. "A Framework for Assessing Ecosystem Dynamics in Response to Chronic Resource Alterations Induced by Global Change." *Ecology* 90: 3279–3289.

Snow, N. 2005. "Successfully Curating Smaller Herbaria and Natural History Collections in Academic Settings." *BioScience* 55: 771–779.

Sokal, R. R., and F. J. Rohlf. 1995. *Biometry,* 3rd edition. New York: W. H. Freeman.

Stankey, G. H., R. N. Clark, and B. T. Borman. 2006. "Learning to Manage a Complex Ecosystem: Adaptive Management and the Northwest Forest Plan." USDA Forest Service, Pacific Northwest Research Station, Research Paper PN-RP-567.

Steinberg, D. K., C. A. Carlson, N. R. Bates, R. J. Johnson, et al. 2001. "Overview of the US JGOFS Bermuda Atlantic Time-series Study (BATS): A Decade-Scale Look at Ocean Biology and Biogeochemistry." *Deep Sea Research Part II: Topical Studies in Oceanography,* 48: 1405–1447.

Steinitz, H. 1970. "A Critical List of Immigrants via the Suez Canal." Pages 64–75 in Biota of the Red Sea and Eastern Mediterranean 1970/1971. Mimeo.

Sterling, T. D. 1959. "Publications Decisions and Their Possible Effects on Inferences Drawn from Tests of Significance—or Vice Versa." *Journal of the American Statistical Association* 54: 30–34.

Stevens, D. L. 1994. "Implementation of a National Monitoring Program." *Journal of Environmental Management* 42: 1–29.

Store, R., and J. Jokimäki. 2003. "A GIS-based Multi-scale Approach to Habitat Suitability Modeling." *Ecological Modelling* 169: 1–15.

Strasser, B. J. 2008. "GenBank—Natural History in the 21st Century?" *Science* 322: 537–538.

Strevens, M. 2004. "The Causal and Unification Accounts of Explanation Unified—Causally." *Noûs* 38: 154–179.

Strickland, D., B. Kielstra, and D. Ryan Norris. 2011. "Experimental Evidence for a Novel Mechanism Driving Variation in Habitat Quality in a Food-Caching Bird." *Oecologia* 167: 943–950.

Studds, C. E., and P. P. Marra. 2005. "Nonbreeding Habitat Occupancy and Population Processes: An Upgraded Experiment with a Migratory Bird." *Ecology* 86: 2380–2385.

Suarez, A. V., and N. D. Tsutsui. 2004. "The Value of Museum Collections for Research and Society." *BioScience* 54: 66–74.

Sullivan, K. A., and A. M. Ellison. 2006. "The Seed Bank of Hemlock Forests: Implications for Forest Regeneration Following Hemlock Decline." *Journal of the Torrey Botanical Society* 133: 393–402.

Suter, G. W. II. 2001. "Applicability of Indicator Monitoring to Ecological Risk Assessment." *Ecological Indicators* 1: 101–112.

Tallis, H., J. Lubchenco, and 238 co-signatories. 2014. "A Call for Inclusive Conservation." *Nature* 515: 27–28.

Taper, M. L., and S. R. Lele, editors. 2004. *The Nature of Scientific Evidence.* Chicago: University of Chicago Press.

Teder, T., M. Moora, E. Roosaluste, K. Zobel, et al. 2007. "Monitoring of Biological Diversity: A Common-Ground Approach." *Conservation Biology* 21: 313–317.

Tessmar-Raible, K., F. Raible, and E. Arboleda. 2011. "Another Place, Another Timer: Marine Species and the Rhythms of Life." *Bioessays* 33: 165–172.

Tewksbury, J. J., J.G.T. Anderson, J. D. Bakker, T. J. Billo, et al. 2014. "Natural History's Place in Science and Society." *BioScience* 64: 300–310.

Thibault, K. M., S.K.M. Ernest, E. P. White, J. H. Brown, et al. 2010. "Long-Term Insights into the Influence of Precipitation on Community Dynamics in Desert Rodents." *Journal of Mammalogy* 91: 787–797.

Thomer, A., G. Vaidya, R. Guralnick, D. Bloom, et al. 2012. "From Documents to Datasets: A MediaWiki-based Method of Annotating and Extracting Species Observations in Century-Old Field Notebooks. *ZooKeys* 209: 235–253.

Tingley, M. W., W. B. Monahan, S. R. Beissinger, and C. Moritz, C. 2009. "Birds Track Their Grinnellian Niche through a Century of Climate Change." *Proceedings of the National Academy of Sciences, USA*, 106: 19637–19643.

Tishbi, A., editor. 2001. *Holy Land in Maps.* Jerusalem: Israel Museum.

Tobler, T. 1849. *El-Kuds, Grundriss von Jerusalem nach Catherwood and Robinson, mit einen neu eingezeichneten Gassennetze und etlichen, theils zur ersten Male erscheinenden, theils berichtigen Gräberplänen.* St. Gallen and Bern, Switzerland: Huber.

——. 1857. *Planographie von Jerusalem.* Gotha, Germany: Justus Perthes.

——. 1867. *Bibliographia geographica Palaestinae: Zunächst kritische Uebersicht gedruckter und ungedruckter Beschreibungen der Reisen ins heilige Land.* Leipzig: S. Hirzel.

Toth, L. A. 2015. "The Influence of the Cage Environment on Rodent Physiology and Behavior: Implications for Reproducibility of Pre-clinical Rodent Research." *Experimental Neurology.* Available at http://dx.doi.org/10.1016/j.expneurol.2015.04.010. Accessed 29 June 2016.

Tottrup, A. P., R. H. Klaassen, M. W. Kristensen, R. Strandberg, et al. 2012. "Drought in Africa Caused Delayed Arrival of European Songbirds." *Science* 338: 1307.

Tsangaras, K., and A. D. Greenwood. 2012. "Museums and Disease: Using Tissue Archive and Museum Samples to Study Pathogens." *Annals of Anatomy* 194: 58–73.

Underwood, A. J. 1997. *Experiments in Ecology: Their Logical Design and Interpretation Using Analysis of Variance.* Cambridge: Cambridge University Press.

Usubutun, S., C. Agalar, C. Diri, and R. Turkyilmaz. 1997. "Single Dose Ciprofloxacin in Cholera." *European Journal of Emergency Medicine* 4: 145–149.

van de Velde, C.W.M. 1858. *Plan of the Town and Environs of Jerusalem Constructed from the English Ordnance-Survey and Measurements of Dr. Titus Tobler, with Memoir by Dr. Titus Tobler.* Gotha, Germany: Justus Perthes.

van Fraassen, B. C. 1980. *The Scientific Image.* Oxford: Clarendon.

Vempati, U. D., C. Chung, C. Mader, A. Koleti, et al. 2014. "Metadata Standard and Data Exchange Specifications to Describe, Model, and Integrate Complex and Diverse High-Throughput Screening Data from the Library of Integrated Network-based Cellular Signatures (LINCS)." *Journal of Biomolecular Screening* 19 (special issue): 803–816.

Visser, M. E., and C. Both. 2005. "Shifts in Phenology Due to Global Climate Change: The Need for a Yardstick." *Proceedings of the Royal Society B: Biological Sciences,* 272: 2561–2569.

Visser, M. E., S. P. Caro, K. van Oers, S. V. Schaper, et al. 2010. "Phenology, Seasonal Timing and Circannual Rhythms: Towards a Unified Framework." *Philosophical Transactions of the Royal Society B: Biological Sciences,* 365: 3113–3127.

Vittoz, P., N. Bayfield, R. Brooker, D. A. Elston, et al. 2010. "Reproducibility of Species Lists, Visual Cover Estimates and Frequency Methods for Recording High-Mountain Vegetation." *Journal of Vegetation Science* 21: 1035–1047.

von Hagen, V.W.F. 1968. *Catherwood, Architect—Explorer of Two Worlds.* Barre, VT: Barre.

von Helmholtz, H. 1847. "Über die Erhaltung der Kraft." *Wisseschaftliche Abhandlungen* 1: 12–75 (reprinted 1882).

von Schubert, G. H. 1839. *Reise in das Morgenland in den Jahren 1836 und 1837.* Erlangen, Germany: J. J. Palm and Ernst Enfe.

Vos, P., E. Meelis, and W. J. ter Keurs. 2000. "A Framework for the Design of Ecological Monitoring Programs as a Tool for Environmental and Nature Management." *Environmental Monitoring and Assessment* 61: 317–344.

Vui, T. Q., J. E. Lohr, P. H. Son, M. N. Kyule, et al. 2004. "Antibody Levels against Newcastle Disease Virus in Chickens in Rural Vietnam." *Proceedings of Control of Newcastle Disease and Duck Plague in Village Poultry* 117: 13–17.

Wallace, C. K., P. N. Anderson, T. C. Brown, S. R. Khanra, et al. 1968. "Optimal Antibiotic Therapy in Cholera." *Bulletin of the World Health Organization* 39: 239–245.

Wallis, J. C., C. L. Borgman, M. S. Mayernik, A. Pepe, et al. 2007. "Know Thy Sensor: Trust, Data Quality, and Data Integrity in Scientific Digital

Libraries." *11th European Conference on Research and Advanced Technology for Digital Libraries, Proceedings,* 4675: 380–391.

Wandeler, P., P.E.A. Hoeck, and L. F. Keller. 2007. "Back to the Future: Museum Specimens in Population Genetics." *Trends in Ecology and Evolution* 22: 634–642.

Warren, D. L., M. Cardillo, D. F. Rosauer, and D. I. Bolnick. 2014. "Mistaking Geography for Biology: Inferring Processes from Species Distributions." *Trends in Ecology and Evolution* 29: 572–580.

Watson, C. M. 1915. *Palestine Exploration Fund: Fifty Years' Work in the Holy Land; A Record and a Summary, 1865–1915.* London: Committee on the Palestine Exploration Fund.

Weiner, A. 1980. "The Function of the Tragic Greek Chorus." *Theatre Journal* 32: 205–212.

Weisberg, M. 2006. "Forty Years of 'the Strategy': Levins on Model Building and Idealization." *Biology and Philosophy* 21: 623–645.

———. 2007. "Three Kinds of Idealization." *Journal of Philosophy* 104: 639–59.

———. 2013. *Simulation and Similarity: Using Models to Understand the World.* Oxford: Oxford University Press.

Wesche, P. L., D. J. Gaffney, and P. D. Keightley. 2004. "DNA Sequence Error Rates in Genbank Records Estimated Using the Mouse Genome as a Reference." *DNA Sequence* 15: 362–364.

Westphal, J. H. 1825. "Ueber die topographische Lage Jerusalems, zur Erläuterung des Planes aufgenommen von dem Herrn Dr. Westphal in Göttingen: Aus dessen Tagebuche während einer Reise durch den Orient in den Jahren 1822 und 1823." *Hertha* 1: 385–390.

White, E. P., E. Baldridge, Z. T. Brym, K. J. Locey, et al. 2013. "Nine Simple Ways to Make It Easier to (Re)Use Your Data." *Ideas in Ecology and Evolution* 6: 1–10. Available at http://dx.doi.org/10.4033/iee.2013.6b.6.f. Accessed 29 June 2016.

Whitlock, M. C. 2011. "Data Archiving in Ecology and Evolution: Best Practices." *Trends in Ecology and Evolution* 26: 61–65.

WHO (World Health Organization). 2009. "Cholera in Zimbabwe—Update 4." Geneva: WHO. Available at http://www.who.int/csr/don/2009_06_09/en/. Accessed 29 June 2016.

———. 2013. "Cholera in Mexico—Update." Geneva: WHO. Available at http://www.who.int/csr/don/2013_10_28/en/. Accessed 29 June 2016.

———. 2014a. "Cholera Outbreak, South Sudan." Geneva: WHO. Available at http://www.who.int/csr/don/2014_05_30/en/. Accessed 29 June 2016.

———. 2014b. "Humanitarian Health Action: Haiti." Geneva: WHO. Available at http://www.who.int/hac/crises/hti/en/. Accessed 29 June 2016.

Wieczorek , J., Q. Guo, and R. J. Hijmans. 2004. "The Point-Radius Method for Georeferencing Locality Descriptions and Calculating Associated Uncertainty." *International Journal of Geographical Information Science* 18: 745–767.

Wielgus, J. 2003. "The Coral Reef of Eilat (Northern Red Sea) Requires Immediate Protection." *Marine Ecology Progress Series* 263: 307.

Wiley, A. E., P. H. Ostrom, A. J. Welch, R. C. Fleischer, et al. 2013. "Millennial-Scale Isotope Records from a Wide-Ranging Predator Show Evidence of Recent Human Impact to Oceanic Food Webs." *Proceedings of the National Academy of Sciences, USA*, 110: 8972–8977.

Williams, G. 1849. *The Holy City—Historical, Topographical and Antiquarian Notices of Jerusalem*. London: John W. Parker.

Williamson, C. J., J. Najorka, R. Perkins, M. L. Yallop, et al. 2014. "Skeletal Mineralogy of Geniculate Corallines: Providing Context for Climate Change and Ocean Acidification Research." *Marine Ecology Progress Series* 513: 71–84.

Wilson, C. W. 1865. *Ordnance Survey of Jerusalem*. London: Her Majesty's Stationery Office.

Wimsatt, W. C. 1980. "Reductionist Research Strategies and Their Biases in the Units of Selection Controversy." Pages 155–202 in *Conceptual Issues in Ecology*, edited by E. Saarinen. Dordrecht: D. Reidel.

———. 2007. *Re-engineering Philosophy for Limited Beings*. Cambridge. MA: Harvard University Press.

Winsor, P. 2003. "Non-essentialist Methods in Pre-Darwinian Taxonomy." *Biology and Philosophy* 18: 387–400.

Wittgenstein, L. 1953. *Philosophical Investigations*, translated by G.E.M. Anscombe. Oxford: Basil Blackwell.

———. 1979. *Remarks on Frazer's Golden Bough*, translated by A. C. Miles. Retford, UK: Brymill.

Woodward, W. 2003. *Making Things Happen: A Theory of Causal Explanation*. New York: Oxford University Press.

Yoccoz, N. G., J. D. Nichols, and T. Boulinier. 2001. "Monitoring of Biological Diversity in Space and Time." *Trends in Ecology and Evolution* 16: 446–453.

Yorks, T. E., D. J. Leopold, and D. J. Raynal. 2003. "Effects of *Tsuga canadensis* Mortality on Soil Water Chemistry and Understory Vegetation: Possible Consequences of an Invasive Insect Herbivore." *Canadian Journal of Forest Research* 33: 1525–1537.

Yosipovich, R., E. Aizenshtein, R. Shadmon, S. Krispel, et al. 2015. "Overcoming the Susceptibility Gap between Maternal Antibody Disappearance and Auto-Antibody Production." *Vaccine* 33: 472–478.

Zalmanovici Trestioreanu, A., H. Green, M. Paul, J. Yaphe, et al. 2010. "Antimicrobial Agents for Treating Uncomplicated Urinary Tract Infection in Women." *Cochrane Database Systematic Reviews* CD007182.

Zanette, L. Y., M. Clinchy, and J. P. Suraci. 2014. "Diagnosing Predation Risk Effects on Demography: Can Measuring Physiology Provide the Means?" *Oecologia* 176: 637–651.

Zantke, J., T. Ishikawa-Fujiwara, E. Arboleda, C. Lohs, et al. 2013. "Circadian and Circalunar Clock Interactions in a Marine Annelid." *Cell Reports* 5: 99–113.

Zhao, X., B. S. Lerner, L. J. Osterweil, E. R. Boose, et al. 2012. "Provenance Support for Rework." In *Proceedings of the 4th USENIX Workshop on the Theory and Practice of Provenance (TaPP '12)*, Cambridge, Massachusetts. Berkeley, CA: USENIX Association. Available at http://dl.acm.org/citation .cfm?id=2342889&CFID=127122717&CFTOKEN=44223646. Accessed 29 June 2016.

Zimmerman, A. S. 2008. "New Knowledge from Old Data—The Role of Standards in the Sharing and Reuse of Ecological Data." *Science Technology and Human Values* 33: 631–652.

CONTRIBUTORS

AYELET SHAVIT is a philosopher of biology, a senior lecturer at Tel Hai College in Israel, and a former Fulbright scholar and Marie Currie (OIF) fellow. Her Ph.D. (summa cum laude) is from the Hebrew University of Jerusalem, and her post-doctoral research was conducted at the Centre for Population Biology and the Department of Philosophy at the University of California, Davis. Her articles appear in journals such as *Philosophy of Science, BioScience,* and *Isis.* Her first book, *One for All?,* was published by The Magnes Press, and her co-edited *Landscapes of Collectivity* is forthcoming from MIT Press.

AARON M. ELLISON is the senior research fellow in ecology at Harvard Forest, Harvard University. He has a B.A. from Yale University in East Asian philosophy and a Ph.D. from Brown University in evolutionary ecology and has a strong interest in conceptual analysis in ecology, particularly on the concept of replication. Aaron was the editor in chief of *Ecological Monographs* (2009–2015) and is a fellow of the Ecological Society of America. He has written over 180 research papers and two books: *A Primer of Ecological Statistics* (Sinauer Associates) and *A Field Guide to the Ants of New England* (Yale University Press).

YEMIMA BEN-MENAHEM is a professor of philosophy and director of the Edelstein Center for the History and Philosophy of Science, Technology, and Medicine at the Hebrew University of Jerusalem. She is the author of *Conventionalism* (Cambridge University Press, 2006), editor of *Hilary Putnam* (Cambridge University Press, 2005), and co-editor of *Probability in Physics* (Springer, 2011). She is currently working on causal constraints, focusing in particular on the interrelations between different causal notions such as determinism and stability, determinism and locality, and so on. Other interests include American pragmatism and the philosophy of history.

NAAMA BERG is the coordinator of Israel's national biodiversity monitoring program at Hamaarag: The Israel National Nature Assessment Program. She holds a Ph.D. in soil ecology from Bar-Ilan University. She led the effort of consolidating the national biodiversity monitoring program and formulating the monitoring protocols. Her main interest is in communicating science to policy makers and ecologically oriented citizen scientists.

DAVID BLANKMAN is the chair of the Information Management Committee of the International Long-Term Ecological Research (ILTER) stations and the former database manager of Israel's LTER Network. He has a master's degree in community and regional planning and has many years of experience and expertise in ecological informatics and database design. His publications appear in journals such as *Ecological Informatics*.

EMERY R. BOOSE is the information manager at Harvard Forest, Harvard University. He has a B.A. in mathematics from Harvard College and a Ph.D. in Sanskrit and Indian studies from Harvard University. Emery oversees management of scientific data and metadata from the Harvard Forest LTER program and often participates in the design and construction of experimental installations in the field. His research interests range from computer science and informatics to meteorology and hydrology, with a recent focus on data provenance and quality control for sensor networks.

AVIGDOR CAHANER is a professor of quantitative genetics in the Faculty of Agriculture at Hebrew University, where he also received his Ph.D. His research, published in more than 100 peer-reviewed articles, focuses on the genetics of immune responses, fat deposition, oxygen supply, meat yield and quality, and heat tolerance of meat-type chickens (broilers). In addition to teaching quantitative genetics, he teaches a course in experimental design and statistical analysis, which is required for all the students in the Faculty of Agriculture.

TAMAR DAYAN is a professor of zoology and director of the National Collections of Natural History at Tel Aviv University. She was a Rothschild, Weizmann, and Fulbright Post-Doctoral Scholar (1989–1991) and received the Bruno Fellowship in 2000. She conducts research in the fields of ecology (evolution of body size and of activity patterns, development of human-animal interactions) and conservation biology (primary drivers of global change: land transformation, climate change, invasive species). She has published 115 papers in peer-reviewed journals and 21 book chapters in peer-reviewed books and has served as a member of the board of directors of Israel's Nature and Parks Authority; chair of the board's Science Committee; chair of Israel's UNESCO Man and the Biosphere Committee; and chair of the board of directors of the Society for the Protection of Nature in Israel (SPNI). Currently she is co-chair of the Forum on Biodiversity of the Environment, operating under the auspices of the Israel Academy of Sciences and Humanities, and chair of the Council of the Open Landscape Institute.

RON DRORI is a remote sensing scientist at Hamaarag: The Israel National Nature Assessment Program. He holds a Ph.D in physical geography from the Hebrew University of Jerusalem. His main interest is remote sensing of vegetation and land-use classification.

BELLA GALIL is an associate professor at Tel Aviv University and a senior scientist at the Israel Oceanographic and Limnological Research Institute in Haifa. Her research interests focus on marine biology with emphases on decapod crustacean taxonomy, dynamics and conservation of marine biodiversity, anthropogenic changes, and marine bioinvasions. She has published over 230 peer-reviewed papers and co-edited three books, and she sits on the editorial boards of three international journals. Bella also is the chairperson of the executive committee of the Zoological Society of Israel; managing editor of the *Israel Journal of Zoology*; founding member of the Israeli Association for Aquatic Sciences (IAAS); member of the Directorate of Israel's Nature and Parks Authority; and chairperson of its scientific committee.

AMATZIA GENIN is a professor of ecology at the Hebrew University of Jerusalem and a consulting associate professor in the Department of Civil and Environmental Engineering, Stanford University. He served as head of the curricular section of ecology, evolution, and behavior at the Hebrew University multiple times and is currently the director of the Interuniversity Institute for Marine Sciences in Eilat. He founded Israel's national monitoring program at the Gulf of Aqaba and is the program's scientific director. His current interests include biological-physical interactions in benthic and pelagic communities, behavior and predator-prey interactions in plankton, flow effects on marine organisms, and coral reef ecology, all subjects on which he has published extensively in leading journals including *Nature, Science*, and *Proceedings of the National Academy of Sciences*.

ORIT GINZBURG holds a Ph.D in ecology from Bar-Ilan University in Israel. Currently she works as the director of rangeland management in Israel's Ministry of Agriculture and Rural Development. Orit participated as an external consultant in a number of environmental projects, including the development of a biodiversity monitoring program in agriculture areas. Her principal interests are agro-environmental policy, natural resources management, conservation, and ecosystem management.

HAIM GOREN is a professor of geography and chairs the program in Galilee Studies at Tel-Hai Academic College, Upper Galilee, Israel. His main fields of interest include Holy Land pilgrims' and travelers' literature, European activity in Ottoman Palestine and the Near East, and the history of the modern scientific study of these regions, currently concentrating on historical geography and cartography. He received his B.A. from the University of Haifa and his M.A. and Ph.D. from the Hebrew University in Jerusalem. He has written or edited more than a dozen books in English, Hebrew, or German and more than 60 refereed papers on the historical geography of Palestine and the Middle East.

BARBARA HELM is an associate professor at the University of Glasgow in the United Kingdom. Her main interests are chronobiology (the study of biological rhythms), ornithology, and animal migration. Prior to her appointment in Glasgow she worked for many years at the Max Planck Institutes for Behavioural Physiology and Ornithology (Germany). Her degree work in biology included an M.Sc. (University of Tuebingen, Germany) and a Ph.D. (Ludwig Maximilian University, Munich, Germany). She has also received an M.A. in philosophy (University of Tuebingen). So far she has written some 70 publications, most of which cover avian chronobiology. Recent interests include the ability of organisms to cope with anthropogenic change, especially effects of changes in seasonality and of light pollution. Since August 2013 she has served as president of the European Ornithologists' Union.

LEONARD LEIBOVICI heads the Department of Medicine at the Rabin Medical Center, Beilinson Campus, Israel, and is a full professor of medicine at the Sackler Faculty of Medicine, Tel Aviv University. His main research interests are computerized decision support, antibiotic treatment, and evidence-based medicine. He has published over 170 articles in peer-reviewed journals, among them the *Journal of the American Medical Association, the British Journal of Medicine* (*BMJ*), and the *Annals of Internal Medicine;* is an editor with the *Journal of Antimicrobial Chemotherapy,* and served on the editorial board of the *BMJ* for five years.

YA'ARA LEIBOVICI-WEISSMAN received her M.D. from the Sackler School of Medicine at Tel Aviv University and recently completed her residency in internal medicine at the Rabin Medical Center in Israel. She has published several studies in the field of infectious diseases.

BARBARA S. LERNER is an associate professor of computer science at Mount Holyoke College. She received her Ph.D. in Computer Science from Carnegie Mellon University. Her recent research has focused on scientific data provenance. This work has been done collaboratively with ecologists, resulting in publications in both computer science and environmental science venues.

MICAL PAUL is the head of the Department of Infectious Disease at the Rambam Hospital, Haifa, Israel, and an associate professor at the Sackler Faculty of Medicine, Tel Aviv University. She graduated from the Hebrew University of Jerusalem and currently is an associate editor with *Clinical Microbiology and Infection* and the Cochrane Infectious Diseases Group. Her research specialty is in the field of infectious diseases and antibiotic management, and she extensively uses meta-analysis in her research. She has published over 80 peer-reviewed publications in journals including *BMJ, Annals of Internal Medicine* and *Clinical Infectious Diseases.*

AVI PEREVOLOTSKY is a professor at the Agricultural Research Organization, Israel. He has a Ph.D. in ecology from the University of California, Davis, and has conducted field research in Peru and the Mt. Sinai region of Egypt and across Israel. He served for 5 years as the chief scientist of the Israeli Nature and Parks authority and for the past 25 years has led the research activities at the Ramat Hanadiv LTER station. The outcome of this activity is summed up in a book, *The Management and Conservation of Mediterranean Ecosystems: Ramat Hanadiv and Beyond.* He has written over 100 research papers and 11 chapters in scientific publications and co-edited *Biodiversity in Drylands: Toward a Unified Framework for Research and Management* (Oxford University Press) and *Ecology: Theory and the Israeli Experience* (Carta, in Hebrew). His principal interest is applied ecology, specifically natural resources management (range management and forestry), conservation, and ecosystem management.

JACOB PITCOVSKI is a professor at Tel Hai College in Israel and head of the Vaccine Development Department at the MIGAL Galilee Research Institute. He received his Ph.D. from Hebrew University and did post-doctoral research at Idaho State University. His research focuses on vaccine development, immune system modulators, and cancer immunotherapy.

REBECCA J. ROWE is an assistant professor in the Department of Natural Resources and the Environment at the University of New Hampshire. She holds an A.B. from Bowdoin College and a Ph.D. from the Committee on Evolutionary Biology at the University of Chicago and was a postdoctoral researcher at the Natural History Museum of Utah, University of Utah. Her research focuses on the impacts of climate change and land use on the distributions and community dynamics of small mammals. Her research is motivated by the need for synthesis across ecological disciplines spanning a range of spatial and temporal scales to provide the framework necessary to better understand and forecast the effects of environmental change. Her work also highlights the use of natural history collections data in ecology and conservation.

EHUD SHAHAR recently submitted his Ph.D. dissertation to the Faculty of Medical Sciences at the Hebrew University of Jerusalem. His dissertation work was supervised by Professors Jacob Pitcovski and Raphael Gorodetsky and describes research on disrupting the immune balance in the microenvironment of cancer tumors and using innate leukocyte activation as an approach for cancer immunotherapy. One article based on his dissertation has already been published, another has been accepted for publication, and Ehud has applied for one patent based on this research.

YONATHAN SHAKED is a geologist by training who made a gradual transition to biological and ecological research and currently manages Israel's National Monitoring Program (NMP) at the Gulf of Aqaba. After completing his Ph.D. at the Hebrew University of Jerusalem he did post-doctoral research at the Institute of Marine and Coastal Sciences at Rutgers University on the evolution and biogeography of symbiosis in marine plankton. Since he returned to the Interuniversity Institute for Marine Sciences in Eilat to lead the NMP, he has continued to publish in peer-reviewed journals on topics ranging from geology to ecology at the Gulf of Aqaba.

MORGAN W. TINGLEY is an assistant professor in the Department of Ecology and Evolutionary Biology at the University of Connecticut. In 2011 he received his Ph.D. in environmental science, policy, and management from the University of California, Berkeley, and in 2004 an M.Sc. in zoology from Oxford University. After finishing his Ph.D., Morgan was a David H. Smith Conservation Research Fellow in the Woodrow Wilson School at Princeton University. His research focuses on using historical data sources to understand the processes underlying long-term ecological change.

KRISTIN VANDERBILT is the information manager at the Sevilleta LTER site in New Mexico and research associate professor at the University of New Mexico. She has a Ph.D. in forest biogeochemistry from Oregon State University. Kristin has a strong interest in the development of an information management system to serve the ILTER Network across disparate local languages, cultures, and infrastructures. She has co-led international workshops on this topic in her role as co-chair of the US LTER Committee and has published on these efforts in the journals *Ecological Informatics* and *Ecosphere*.

INDEX

The letter *t* following a page number denotes a table, and the letter *f* denotes a figure. A page span may contain more than one table or figure.